CROSS-LAYER RESOURCE ALLOCATION
IN WIRELESS COMMUNICATIONS:
TECHNIQUES AND MODELS FROM PHY
AND MAC LAYER INTERACTION

Cross-Layer Resource Allocation in Wireless Communications

Techniques and Models from PHY and MAC Layer Interaction

Ana I Pérez-Neira and Marc Realp Campalans

AMSTERDAM • BOSTON • HEIDELBERG • LONDON
NEW YORK • OXFORD • PARIS • SAN DIEGO
SAN FRANCISCO • SINGAPORE • SYDNEY • TOKYO

Academic Press is an imprint of Elsevier

Academic Press is an imprint of Elsevier
Linacre House, Jordan Hill, Oxford, OX2 8DP
84 Theobald's Road, London WC1X 8RR, UK
30 Corporate Drive, Burlington, MA 01803
525 B Street, Suite 1900, San Diego, California 92101-4495, USA

First edition 2009

British Library Cataloguing in Publication Data
Perez-Niera, Ana I.
    Cross-layer resource allocation in wireless communications: techniques and models
    from PHY and MAC layer interaction 1.   Broadband communication systems –
    Standards 2.   Mobile communication systems – Standards 3.   Cellular telephone
    systems – Standards 4.   Resource allocation
    I.   Title II.   Campalans, Marc Realp
    621.3'8456

ISBN-13: 978-0-12-374141-7

Library of Congress Catalog Number: 2008930971

ISBN: 978-0-12-374141-7

For information on all Academic Press Publications
visit our web site at books.elsevier.com

Typeset by Charon Tec Ltd., A Macmillan Company
(www.macmillansolutions.com)

Printed and bound in Great Britain

09 10 11   10 9 8 7 6 5 4 3 2 1

Contents

Preface

Recently, there has been considerable interest in the idea of cross-layer design of wireless networks. This is motivated by the need to provide a greater level of adaptivity to variations of wireless channels. This book examines the interaction between the physical and medium access control layers. In particular, the book considers the impact of signal processing techniques that enable mulipacket transmission and reception on the throughput and design of protocols. Main emphasis is given to the spatial dimension. The book can be found interesting for researchers and professionals working either in the PHY layer or in the MAC layer, who want to get initiated into the MAC or PHY layer, respectively. Concerning resource allocation strategies in wireless communication systems, contributions from the wireless communications research community are either in the information theory field or in the networking field with an evident isolation between them. This book explores the advantages of breaking down such traditional isolation and considers resource allocation cross-layer techniques, models and methodologies that will help researchers from these two fields to increase their synergy.

The book is intended as a reference book for researchers and graduate students of the wireless communications community, which provides with a general framework for cross-layer design. There are many publications and books regarding cross-layer designs, but there is a lack of a general framework. The general framework in this book aims at the joint design of scheduling, power control, adaptive modulation and its interplay with channel state information.

The whole book consists of eight chapters. First, in Chapter 1, the fundamental concepts of this book are introduced. In Chapters 2 and 3, a detailed description of the concept of spectral efficiency in both single-user and multi-user systems is presented. Then, with the performance metric clearly defined and understood, the optimal resource allocation in multi-user SISO systems is studied in Chapter 4. In Chapter 5, multi-user SIMO channels are examined, whereas the multi-user MISO channel is analyzed in Chapter 6. The delay, the other performance metric mentioned previously, is presented in Chapter 7 where the resource allocation strategies presented in previous chapters are analyzed in terms of delay. Finally, the book considers a general perspective on how cross-layer resource allocation should be considered in multi-user OFDMA systems.

Acknowledgements

The authors gratefully acknowledge valuable discussions on the topic with people of CTTC (Centre Tecnològic de Telecomunicacions de Catalunya) and also thank Prof. Lang Tong, Velio Tralli, the reviewers and people at Elsevier who have helped bring the project into reality. The authors would also like to thank the European Commission under project NEWCOM++ (216715), the Spanish Government under projects TEC2005-08122-C03 and PROFIT FIT-330225-2007-2, and the Catalan Government under grant 2005SGR-00996 for their support of much of the research described in this book.

List of Figures

List of Tables

List of Acronyms

2G	Second Generation
3G	Third Generation
AP	Access Point
AGWN	Additive Gaussian White Noise
ARQ	Automatic Repeat Request
BER	Bit Error Rate
BPSK	Binary Phase Shift Keying
BS	Base Station
CAC	Connection Admission Control
CDMA	Code Division Multiple Access
CSI	Channel State Information
DPC	Dirty Paper Coding
FD	Frame Division
FDMA	Frame Division Multiple Access
FEC	Forward Error Correction
GSM	Global System for Mobile communications
HMUD	Heterogeneous Multi User Diversity
HSDPA	High Speed Downlink Packet Access
IFFT	Inverse Fast Fourier Transform
IP	Internet Protocol
IS-95	Interim Standard 95
KKT	Karush–Kuhn–Tucker
LAN	Local Area Network
LDPC	Low-Density Parity-Check code
MAC	Medium Access Control
MCS	Modulation and Coding Scheme
MIMO	Multiple-Input Multiple-Output
MISO	Multiple-Input Single-Output
MMSE	Mininum Mean Square Error
MUD	Multi-User Diversity
MUX	Multiplexing
OSI	Open Systems Interconnection
OFDM	Orthogonal Frequency Division Multiplexing
OFDMA	Orthogonal Frequency Division Multiple Access

PER	Packet Error Rate
PF	Proportional Fair
PHY	PHYsical layer
PSR	Packet Success Rate
PSK	Phase Shift Keying
QAM	Quadrature Amplitude Modulation
QoS	Quality of Service
RR	Round Robin
SIC	Successive Interference Cancelation
SIMO	Single-Input-Multiple-Output
SISO	Single-Input-Single-Output
SER	Symbol Error Rate
SD	Spatial Diversity
SDMA	Spatial Diversity Multiple Access
SNR	Signal to Noise Ratio
SVD	Singular Value Decomposition
TDD	Time Division Duplexing
TDMA	Time Division Multiple Access
UMTS	Universal Mobile Telecommunications System
VoIP	Voice over IP
ZF	Zero Forcing

1

Introduction

The wireless communications industry has experienced a tremendous transformation since the first digital 2G (GSM, IS-95) cellular systems for voice communications which appeared in the mid-1990s. In fact, the current 3G (UMTS, CDMA-2000), the beyond 3G (HSDPA), the wireless LAN (IEEE 802.11x) and the WiMAX (IEEE802.16) systems are destined not only for voice communications but also for the transmission of data. Because of the potential of wireless networks to enable more and more exciting applications, wireless communications have become one of the main areas of research in recent decades.

One of the main characteristics in wireless communications, when compared to their wireline counterpart, is that wireless communications suffer from a very hostile channel of transmission, where the wireless propagation medium is random with time-varying attenuation and multipath, so that realizing reliable and efficient communications is a challenging task. For this reason, much work has mainly concentrated on optimizing the physical (PHY) layer in the communication protocol stack. Also, managing the scarcity of resources in a widely varying environment is also a very serious challenge and, hence, not only is the optimization of the PHY layer important but also other layers of the protocol stack must be taken into account. Recent results [1–3] have shown that there is a strong interconnection among all layers in a wireless network and, thus, a local layered design approach may not be optimal but a cross-layer design might be more efficient.

1.1 The need for a general framework for cross-layer design in wireless systems

Typically, communication systems can be modeled following the Open Systems Interconnection (OSI) standard where different functionalities of the system are distributed into layers. For instance, the PHY layer is charged with guaranteeing a reliable and efficient delivery of bits; this means performing tasks like data modulation and channel coding. The MAC/Link layer is charged with allocating resources to multiple users in the system, which translates to performing multiplexing and scheduling tasks by handling QoS to some extent. Another example is the network layer that has, among others, routing functionalities.

p0040   In this book, we consider the use of an integrated or cross-layer design approach to optimize the performance of wireless networks. Specifically, we focus on the joint design of the MAC/Link and PHY layers. The traditional design paradigm emphasizes layer transparency, which means that a layer does not need to know the inner workings of the layer below or above. A layer only needs to know which functions the layers below and above provide. This is beneficial for many reasons. For instance, the physical layer can be changed if the communication medium (e.g. from copper-line to fiber) is changed, without having to modify the upper layers. While this philosophy is advantageous from a design perspective, it may lead to inferior performance in the wireless transmission.

p0050   One of the pioneering works on cross-layer design in wireless communications is the paper by Knopp and Humblet [4] where the uplink of a single cell with users experiencing fast fading channels is considered. The phenomenon of multiple users experiencing independent fading channels is known as multi-user diversity (MUD). In this context, it turns out that the sum capacity (sum of the simultaneous user capacities) is maximized if, for each time instant, the user with the best channel gain is scheduled. The gain achieved with such strategy is known as MUD gain. The optimal solution also includes a power-control law which uses more transmit power for strong channels than for weak channels. This optimal solution is an opposite strategy to conventional power control, which uses transmit power to compensate for weak channels. Similarly, the sum capacity achieving solution for the corresponding downlink scenario is such that, at each time instant, the base station should transmit to the user with the strongest channel. With the informative theoretic results that brought forth MUD, the necessity to co-design parts of the PHY and MAC/Link layers in order to increase performance is clear.

In general terms, contributions for PHY–MAC/Link resource allocation strategies come either from the information theory field [5–8] or from the networking field [9] with an evident isolation between them. This book explores the advantages of breaking down such traditional isolation and considers resource allocation cross-layer techniques, models and methodologies that will help researchers from these two fields to increase their synergy. It will also solve the challenging problem of supporting multimedia applications and services over wireless networks with their constraints and heterogeneities such as limited battery power, limited bandwidth, random time-varying fading effect, different protocols and standards and stringent quality of service (QoS) requirements. In this book, we propose a cross-layer optimization framework to jointly design the scheduling and power control in wireless networks. We study the system performance by combining scheduling, power control, adaptive modulation and its interplay with channel state information (CSI).

1.2 Measuring performance in cross-layer design

For the joint optimization of the PHY and the MAC/Link layers, it is necessary to understand and relate their terms and concepts. Further, the design of a resource allocation policy depends on the definition of a performance measure.

1.2.1 The spectral efficiency

Some common performance measures in the information-theory field are the instantaneous capacity and the ergodic capacity. However, capacity is a theoretical upper bound that cannot be achieved in practice and a practical view of the capacity is necessary when cross-layer resource allocation is to be performed with a more practical perspective and considering suboptimal transceiver architectures and parameters such as channel coding, data modulation and link adaptation. In this book, we identify the concept of spectral efficiency denoted by $R_k(\mathbf{H}, \mathbf{p}(\mathbf{H}))$ as the practical view of the capacity. The spectral efficiency concept depends on the current channel state $\mathbf{H}$ and the resource allocation policy $\mathbf{p}(\mathbf{H})$ and is defined at both the PHY layer and the MAC/Link layer of the communications system so that an accurate cross-layer analysis can be performed.

p0090 In general terms, in multi-user systems with bursty sources, it is not straightforward to define what is meant by good system performance, as this term can refer to many different aspects. The performance metrics can be considered from a network perspective to determine how efficient the network resource is utilized or from a user perspective to determine fairness among users as well as quality of service (QoS), which are also very important measures because they determine how smooth end-user applications are run on the wireless systems.

p0100 Starting from a general formulation, one of the aims of this book is to show that performance can be defined in terms of the utility function $\sum_k \theta_k R_k(\mathbf{H}, \mathbf{p}(\mathbf{H}))$, from which the resource allocation policy is $\mathbf{p}(\mathbf{H})$ obtained. Then, the business model of the operator influences the desirable behavior of the resource allocation policy through the design of the different user weights θ_k. It is found that in many situations, the sum of weighted spectral efficiencies is convex with respect to $\mathbf{p}(\mathbf{H})$ so that convex optimization methods can be used. Unfortunately, obtaining the user weights in order to achieve a desired system performance is not always an easy task.

p0110 Alternatively, some similar heuristic resource allocation approaches exist in the literature. Recent studies on user satisfaction indicate that the satisfaction of an already 'well-served' user increases only marginally by increasing the service level even further. However, if the service level is decreased below some level, the satisfaction level drops significantly. One popular utility function used to strike a balance between system spectral efficiency and fairness among users is called proportional fairness (PF). A resource allocation policy is called proportional fair if it optimizes $\sum_k \log_2(\bar{R}_k)$ where $\bar{R}_k$ is the average spectral efficiency of user k. In this function, there is a heavy penalty for terms with small $\bar{R}_k$ due to the concavity of the $\log_2(.)$. Hence, to maximize the utility, the scheduler has to avoid the situation where some users obtain very small spectral efficiency. Therefore, the logarithmic utility function connects user satisfaction with the QoS. A scheduler that maximizes $\sum_k \frac{R_k(\mathbf{H}, \mathbf{p}(\mathbf{H}))}{\bar{R}_k}$ will also maximize the PF utility function. Recently, variations of PF have been proposed [10]. This book will provide some insight into the relationship between the optimal resource allocation strategy which maximizes $\sum_k \theta_k R_k(\mathbf{H}, \mathbf{p}(\mathbf{H}))$ with heuristic resource allocation strategies such as the PF.

For the sake of completeness, it is worth commenting that the previously presented utility functions do not give hard QoS guarantee [11, 12]. For that purpose the alternative is to optimize an instantaneous cost in terms of users, power and/or modulation subject to instantaneous or outage constraints [13], at the expense of lower system performance [14]. This setting reflects a sophisticated problem of resource allocation combining an efficiency objective with strict QoS constraints. In most of the cases, the problem can be formulated also as a convex optimization problem and unlike the case of weighted sum spectral efficiency optimization, the problem of feasibility arises [15, 16].

1.2.2 The delay

From an information theory point of view, the ergodic capacity can be achieved without instantaneous feedback of the channel but with a capacity-achieving codeword that is long enough so the transmitted data is encoded over all the possible channel fading states. Although this can be easily assumed in rapid varying (or fast fading) channels, this is not necessarily true in slow fading channels where delay would be very high. Nevertheless, short codes could be used in slow fading channels and, hence, delay could be reduced at the expense of experiencing transmission errors. Errors occur when the SNR is below a threshold determined by the code. In that case, we say that the system is in outage. Besides, the probability of the SNR being below the threshold is the outage probability. Clearly, the shorter the code, the lower the delay but the higher the outage probability.

A similar approach can be used from a network point of view because, in general, the higher the transmission rate, the higher the probability of error but the lower the delay. However, not only do PHY layer parameters such as the channel coding and data modulation affect the delay but, in multi-user communications, resource allocation and data storage are important aspects in terms of delay.

The fact that delay requirements should be treated differently in different layers suggests a cross-layer approach for delay analysis.

In fact, the cross-layer approach has been applied to the design and analysis of QoS-featured multi-access systems by a few researchers recently. For example, the analysis of queueing delay for 802.16 networks was conducted in [17–19] by combining link-layer queueing with physical-layer transmission. A vacation queueing model was adopted in [20] to analyze the link-layer queueing performance of OFDM–TDMA systems with round-robin scheduling. A queueing model for OFDMA systems was used in [21] to design a scheduling scheme that balances multi-user diversity and queueing delay. In [22], the authors facilitate the efficient support of QoS by modeling a wireless channel in terms of connection-level QoS metrics such as data rate, delay and delay-violation probability. In this book, we relate the spectral efficiency measure to different measures of delay performance such as queueing delay and access delay.

s0050

1.3 Considering multiple antennas

p0160

In multi-user systems, besides the MUD gain, when terminals are provided with multiple antennas, the spatial diversity (SD) gain and the multiplexing (MUX) gain also play an important role. Indeed, multiple antennas can be used to increase the user's SNR (SD gain) or to increase the number of streams transmitting simultaneously (MUX gain). Resource allocation in MIMO systems aims at finding the optimal combination of MUD, SD and MUX gains by dynamically scheduling terminals' powers in space and time.

p0170

From an information-theory perspective the fundamental limits of the wireless MIMO channel are clear [23, 24]. However, the optimal transceiver architecture that achieves the ergodic sum capacity in both the multiple access channel and the broadcast channel is rather complex and, hence, suboptimal beamforming architectures might be considered. In that case, the problem of maximizing the weighted sum of spectral efficiencies for a given architecture is not always convex due to the intricate relationship of the interferences [25].

p0180

By means of a zero forcing (ZF) beamformer, interferences are nulled at the expense of increasing the noise power and, hence, the optimization problem becomes convex. For other non-orthogonal beamforming schemes the ability of the beamforming design to perform well depends highly on the users allocated with some power. The more spatially separated the users are, the higher the performance that can be expected. Therefore, in many new MIMO schemes, the user selection and beamforming design has merged into one, usually highly complex, block. Unfortunately, the design of the optimal beamformer is a non-convex optimization problem. For the sake of convexity, the ZF architecture will be considered as the reference beamformer in this book. In this way, the optimal power allocation strategies in the multiple access channel and in the broadcast channel are obtained.

190 The design and operation of the optimal power allocation depends highly on how much CSI is available at the transmitter, and in which form. By considering a ZF beamformer, the optimal solution relies on the use of complete CSI at the base station which might be a computationally complex costly solution. First, because the need of instantaneous and complete CSI imposes the design of specific signaling channels in order to estimate the channel state of all users even when only a reduced set of users will be scheduled. And second, because optimal policies typically imply an exhaustive search over all the possible sets of users. In the case of finite-rate feedback links from the users, the CSI can only be partial. In this book, we also study the cross-layer design of low-rate feedback resource allocation strategies, in particular for multi-user diversity and multi-antenna systems. Multi-user MIMO systems in practice reduce to either multi-user MISO or SIMO systems, and this is the focus of this book, otherwise the required feedback makes the whole multi-user system a non-realistic one.

060

1.4 Considering Orthogonal Frequency Division Multiple Access (OFDMA)

0200 Since the frequency-selective nature of the wireless channel imposes some problems on broadband systems that rely on conventional single-carrier techniques, more and more wireless devices are based on OFDMA. As OFDMA systems provide excellent physical layer properties, they also offer interesting opportunities regarding link-layer aspects. Due to the relatively fine granularity of the subchannels, resource requirements of terminals can be served, in principle, without much overprovisioning of bandwidth. In addition, due to the diversity of such systems (in frequency, time and space), the modulation type and the transmit power per subchannel can be adapted in order to increase spectral efficiency. In a multi-user OFDMA system, diversity can be exploited by dynamically assigning different sets of subcarriers to different terminals. Cross-layer optimization approaches attempt to dynamically match the requirements of data link connections to the instantaneous physical layer resources available in order to maximize some system metric. In [26] the authors review a few representative basic approaches for point-to-point and point-to-multi-point communications, which serve as design references for future system concepts.

1.5 Book structure

The general block diagram of the system we are studying in this book is shown in Figure 1.1. Either in the uplink or in the downlink, a resource allocation unit located at the base station (BS) or access point (AP) is charged with spatially multiplexing and scheduling users by means of performing power allocation to users. After allocating power to a set of users $\{1, \dots, K\}$, because some users might have zero power allocated to them, a subset of active users K with cardinality $|K|$ is spatially multiplexed and scheduled.

Then, in the uplink, these users transmit data using a given modulation and coding scheme (MCS) according to a link adaptation procedure and the users' signals are processed at the receiver by a beamformer (typically a zero forcing beamformer). In the downlink, the process is similar; the optimal MCS is chosen according to a link adaptation process and users' signals are processed by a beamformer before being transmitted.

Having such general structure in mind, Chapters 2 and 3 of the book present a detailed description of the concept of spectral efficiency in both single-user and multi-user systems, respectively. Then, with the performance metric clearly defined and understood, the optimal resource allocation in multi-user SISO systems is studied in Chapter 4. In Chapter 5, multi-user SIMO channels are examined, whereas the multi-user MISO channel is analyzed in Chapter 6. The delay, the other performance metric mentioned previously, is

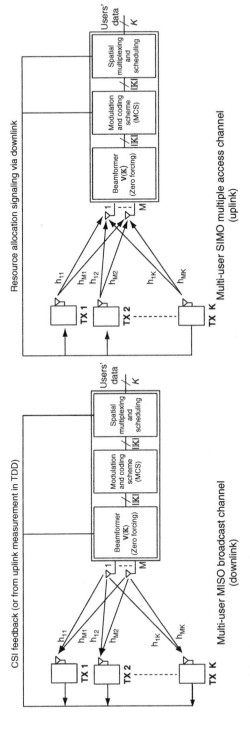

f0010 **Figure 1.1** *General system architecture.*

presented in Chapter 7 where the resource allocation strategies presented in previous chapters are analyzed in terms of delay. Finally, the book considers a general perspective on how cross-layer resource allocation should be considered in multi-user OFDMA systems.

References

[1] *IEEE Journal on Selected Areas in Communications*, Vol. 25, No. 4, May 2007.

[2] *Eurasip Journal on Advances in Signal Processing*, Special issue on Cross-layer Design for the Physical, MAC, and Link Layer in Wireless Systems, Aug. 2008.

[3] *IEEE Signal Processing*, Special issue on Advances in Signal Processing-assisted Cross-layer Designs, Vol. 86, Issue 8, Aug. 2006.

[4] R. Knopp and P.A. Humblet, Information capacity and power control in single-cell multiuser communications, *Proc. IEEE ICC.*, June 1995.

[5] G.B. Giannakis, Y. Hua, P. Stoica and L. Tong, *Signal Processing Advances in Wireless and Mobile Communications: Trends in Single- and Multi-user Systems,* Prentice Hall PTR, 2001.

[6] C. Comaniciu, N.B. Mandayam and H.V. Poor, *Wireless Networks: Multiuser Detection in Cross-layer Design,* Springer, 2005.

[7] W.L. Huang and K.B. Letaief, Cross-layer scheduling and power control combined with adaptive modulation for wireless ad-hoc networks, *IEEE Trans. Comm*, Vol. 55, Issue 4, pp. 728–739, April 2007.

[8] J. Lee, R.V. Sonalkar and J.M. Cioffi, Multiuser bit loading for multicarrier systems, *IEEE Trans. Comm.*, Vol. 54, Issue 7, pp. 1170–1174, July 2006.

[9] L. Georgiadis, M.J. Neely and L. Tassiulas, *Resource Allocation and Cross Layer Control in Wireless Networks,* Now Publishers Inc., 2006.

[10] P. Svedman, Cross-layer aspects in OFDMA scheduling, PhD dissertation, June 2007.

[11] Y.-C. Liang, R. Zhang and J.M. Cioffi, Transmit optimization for MIMO-OFDM with delay-constrained and

no-delay-constrained traffic, *IEEE Trans. Signal Processing.*, Vol. 54, Issue 8, pp. 3190–3199, Aug. 2006.

[12] D. Liao, L. Li, S. Xu and H. Yu, Opportunistic scheduling with multiple QoS constraints in wireless multiservice networks, *Proc. IEEE WCNC, 2007*, March 2007.

[13] N. Zorba and A. Pérez-Neira, Robust multibeam opportunistic schemes under quality of service constraints, *Proc. ICC, 2007*, May 2007.

[14] N. Zorba and A. Pérez-Neira, Optimum number of beams in multiuser opportunistic schemes under QoS constraints, *Proc. International ITG/IEEE Workshop on Smart Antennas TechGate*, Feb. 2007.

[15] T. Michel and G. Wunder, Achieving QoS and efficiency in the MIMO downlink with limited power, *Proc. International ITG/IEEE Workshop on Smart Antennas TechGate*, Feb. 2007.

[16] D. Liao, L. Li, S. Xu and H. Yu, Traffic aided opportunistic scheduling with QoS support for multiservice CDMA uplink, *Proc. IEEE WCNC, 2007*, March 2007.

[17] R. Iyengar, et al., Delay analysis of 802.16 based last mile wireless networks, *Proc. IEEE Globecom, 2005*, Nov. 2005.

[18] D. Niyato, et al., Queue-aware uplink bandwidth allocation for polling services in 802.16 broadband wireless networks, *Proc. IEEE Globecom, 2005*, Nov. 2005.

[19] D. Rajan, A. Sabharwal and B. Aazhang, Delay-bounded packet scheduling of bursty traffic over wireless channels, *IEEE Trans. Information Theory*, Vol. 50, Issue 1, pp. 125–144, Jan. 2004.

[20] D. Niyato and E. Hossain, Queueing analysis of OFDM/TDMA systems, *Proc. IEEE Globecom, 2005*, Nov. 2005.

[21] G. Song, Y. Li, L.J. Cimini and H. Zheng, Joint channel-aware and queue-aware data scheduling in multiple shared wireless channels, *Proc. IEEE WCNC, 2004*.

[22] D. Wu and R. Negi, Effective capacity: a wireless link model for support of QoS, *IEEE Trans. Wireless Comm.*, Vol. 2, Issue 4, July 2003.

[23] W. Rhee and J.M. Cioffi, On the capacity of multiuser wireless systems with multiple antennas, *IEEE Trans. Information Theory*, Vol. 49, pp. 2580–2595, Oct. 2003.

[24] H. Weingarten, Y. Steinberg and S. Shamai, The capacity region of the Gaussian multiple-input multiple-output broadcast channel, *IEEE Trans. Information Theory*, Vol. 52, pp. 3936–3964, 2006.

[25] M. Schubert, H. Boche and S. Stanczak, Strict convexity of the QoS feasible region for log-convex interference functions, *Proc. ACSSC'06*.

[26] K.B. Letaief and Y.J. Zhang, Dynamic multiuser resource allocation and adaptation for wireless systems, *IEEE Wireless Comm.*, Special issue on Smart Antennas, Vol. 13, Issue 4, Aug. 2006.

2

Different views of spectral efficiency*

*This chapter is co-authered by Dr. Stephen Pfletschinger

The performance of a wireless data communication system might be measured in many different ways. One key parameter is the spectral efficiency. Spectral efficiency is defined as the amount of error-free information that can be transmitted over a given bandwidth in a specific digital communication system. It is in fact a measure of how efficiently bandwidth resources are used by the different protocols of the communication stack. It is usually measured in bits per second per hertz denoted by bps/Hz if a continuous time model is considered, or similarly in bits per channel use if an equivalent discrete time model is considered instead. That is, the rate at which data can be transmitted error free every time the transmitter gets access to the channel.

In many wireless systems, the communication channel of a transmitter–receiver link scatters the transmitted signal along its propagation path. Channel conditions change in time creating random fluctuations of the received power level, or fading. In order to ensure reliable communications, redundancy is introduced at the transmitter by encoding data before transmission. The ultimate spectral efficiency limit over a channel was given by Shannon in 1948 and is called the channel capacity. However, the channel capacity provides an information theory upper bound of the spectral efficiency based on ideal channel coding with infinite codeword lengths and ideal data modulation schemes. Indeed, the *real* spectral efficiency might be more or less close to channel capacity depending on the channel coding and data modulation schemes used at the PHY layer of the data communication system.

13

p0030　Not only the PHY layer but also the MAC/Link layer has an important effect on the real spectral efficiency. At the MAC/Link layer, data are encapsulated in packets of bits and sent through the wireless channel after being processed by the PHY layer. Due to the wireless and broadcast nature of the channel, such packets might suffer from errors. Then, we define the spectral efficiency at the MAC/Link layer as the amount of error-free data that is delivered to the upper layers of the communication system. We name this measure of spectral efficiency as throughput.

p0040　Let us assume a wireless point-to-point link with one transmitter and one receiver communicating through an additive white Gaussian noise (AWGN) channel. The simplified system model is depicted in Figure 2.1. The channel is modeled as the discrete-time equivalent baseband channel with complex-valued inputs. The fading coefficient is set to $h = 1$ unless otherwise stated and the noise is white AWGN, i.e. $w \approx CN(0, \sigma^2)$.

p0050　In this simplified model, the channel output is given as

$$r = h \cdot s + w \qquad (2.1)$$

p0060　Following Figure 2.1, at the PHY layer of the transmitter, data arrive from the MAC/Link layer encapsulated into packets of bits. Such bits, represented by the vector **u**, are encoded and interleaved forming vectors of bits denoted by **x**. Although not essential, an interleaver Π usually exists to protect the transmission against burst errors. Encoded and interleaved bits are then modulated into symbols s and finally transmitted through the wireless channel. In order to estimate the transmitted bits, the opposite operations are

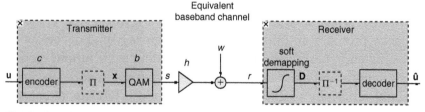

f0010　**Figure 2.1** *Discrete-time system model for evaluation of modulation and coding scheme.*

performed at the receiver side. Note that although explained separately, in practice, coding and modulation operations can be performed jointly or separately.

2.1 The capacity

Let us assume a wireless point-to-point link with one transmitter and one receiver communicating through an AWGN channel. Given an SNR at the receiver denoted by γ and if we consider the equivalent baseband discrete time channel of Figure 2.1, the channel capacity is given by [1]

$$C(\gamma) = \log_2(1 + \gamma) \qquad (2.2)$$

in bits per channel use.

In a MIMO communication link with many antennas at the transmitter and receiver sides, each selection of a pair of transmit and receive antennas gives rise to a different value of the SNR. For that reason, (2.2) is commonly used with γ being defined as the post-detection SNR, i.e. the SNR after the signals from different antennas are combined and processed. However, being rigorous, the concept of channel capacity is defined without considering the use of any kind of transmission/reception scheme and thus, if γ is defined as the post-detection SNR, expression (2.2) cannot be referred to as the channel capacity. For the rest of the book, when we use γ as the post-detection SNR, we will refer to (2.2) as information theory rate or simply rate when no confusion is possible.

2.2 Digital data modulation

Consider the communication system model described in Figure 2.1. The main objective of the digital data modulation block in Figure 2.1 is to map the sequences of bits $\mathbf{x} = \{x_1, \ldots, x_b\} \in X$ into symbols s, where $S = \{s_1, \ldots, s_M\}$ is the set of all possible symbols with $M = 2^b$. Although the communication system model presented in Figure 2.1 is a discrete (digital) one, in practice, symbols s_n are embedded into analog signals that are transmitted through the channel. Since there are $M = 2^b$ sequences of b bits, the system sends $b = \log_2 M$ bits of information per symbol.

s0030
2.3 The bit error rate (BER)

p0100
An important reference for the assessment of any modulation scheme is the bit error probability or bit error rate (BER) for the corresponding uncoded system. Unfortunately, for most non-binary modulation techniques (e.g. M-QAM and M-PSK) an exact expression for BER is hard to find.

p0110
At high SNR and using Gray mapping [2], it is commonly assumed that an erroneous detected symbol differs from the correct one in only one bit. Consequently, the BER is approximated by the symbol error rate (SER) divided by the number of bits per symbol b.

p0120
Closed-form expressions for SER of M-QAM and M-PSK as functions of the SNR can be found in [2]. For M-QAM with square constellations, i.e. b is an even integer, the BER approximation is given by

$$BER_{MQAM}(\gamma) = \frac{2}{\log_2 M}\left(1 - \frac{1}{\sqrt{M}}\right)erfc\left(\sqrt{\frac{3\gamma}{2(M-1)}}\right) \quad (2.3)$$

where $erfc(.)$ is the complementary error function. For M-PSK modulations the BER approximation is

$$BER_{MPSK}(\gamma) = \begin{cases} erfc\left(\sqrt{\gamma}\right) & for\ \log_2 M = 1,2 \\ \dfrac{1}{\log_2 M}erfc\left(\sqrt{\gamma}\sin\left(\dfrac{\pi}{M}\right)\right) & for\ \log_2 M > 2 \end{cases}$$

$$(2.4)$$

where BPSK and QPSK have the same BER because a QPSK signal can be seen as two independent BPSK signals.

p0130
An example is given in Figure 2.2. We observe that for modulations higher than 8-PSK it is preferable to move to QAM modulations. Note that 2-QAM and 4-QAM modulations are equivalent to BPSK and QPSK modulations, respectively. Furthermore, it can be shown that the BER performance of 8-QAM is very close to that of 16-QAM but with one bit per symbol less. Hence, it is quite usual that in commercial systems QAM modulations start at

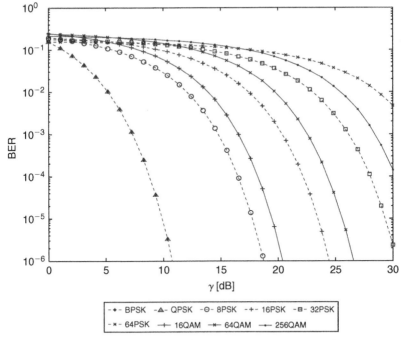

Figure 2.2 *BER curves for different modulations.*

16-QAM. For instance, 16-QAM and 64-QAM are the two QAM schemes considered by the IEEE802.11g/a standards [3].

2.4 Channel coding

At the PHY layer of the receiver side, received symbols are detected, demapped and decoded into estimated bits that are passed to the upper layers of the communication system. During the transmission, symbols are corrupted by different sources such as the wireless channel variability, the noise at the receiver and the interference from other transmitters. Therefore, the estimated bits at reception might be different from the transmitted ones. Such differences are called transmission errors.

Channel coding (encoding and decoding) techniques improve transmission reliability by introducing redundancy to the information sequence that is to be transmitted. Traditionally, channel coding schemes aimed at reducing transmission errors are referred to error correcting strategies. This concept is based on the idea that symbols are hard detected and demapped at the receiver well before they are passed to the decoder. Then, the job of the decoder is to detect and correct possible errors of the incoming bit sequence. However, this has been demonstrated to be a sub-optimal receiver implementation but soft detecting and demapping symbols are shown to offer better performances. In that case, sufficient statistics are provided to the decoder in charge of deciding the sequence of bits that have been transmitted. Clearly, it does not make sense to talk of error correction in this case, because is the decoder the one that introduces errors in the estimation of the transmitted bit sequence? Despite whether hard or soft detection is used, channel codes designed to minimize the errors between the transmitted bits and the estimated bits at reception are named forward error correction (FEC) codes.

p0160   Following Figure 2.1, the encoder is in charge of setting a one-to-one correspondence between input sequences $\mathbf{u}$ of k bits and output sequences $\mathbf{x}$ of n bits (omitting the interleaver). The set of all possible sequences $\mathbf{x}$ is the channel code and the ratio $c = k/n$ is called the code rate. The sequences $\mathbf{u}$ at the input of the encoder are called data words and the sequences $\mathbf{x}$ at the output of the encoder are called code words.

p0170   Many decades of research have been dedicated to the search for channel coding schemes that come close to fulfilling the promise of Shannon's theorem. These activities produced a rich body of theory on block and convolutional codes until the invention of turbo codes [21] brought the breakthrough in achieving performance within fraction of dBs from the Shannon limit. This innovation led to the rediscovery of LDPC codes [19, 20], which share with turbo codes the possibility for iterative decoding. A modern overview of coding theory can be found in [17, 18].

s0050   ## 2.5 The packet error rate (PER)

p0180   In practical systems, many data modulations are combined with different coding rates. Therefore, we use m to denote the Modulation and Coding Scheme (MCS) defined by the tuple $(b^{(m)}, c^{(m)})$ corresponding to a modulation of $b^{(m)}$ bits per symbol and a coding rate $c^{(m)}$. Given an MCS m, the number of information bits transmitted per channel use is $b^{(m)}c^{(m)}$. Let us name

$$R^{(m)} = b^{(m)}c^{(m)} \qquad (2.5)$$

as the effective transmission rate.

At the receiver side, the received signal is demodulated and bits are decoded and passed to the MAC/Link layer on a packet-by-packet basis. One or many data words might conform a single packet. However, for simplicity reasons and without loss of generality, we will assume for the rest of the book that one packet is composed of one data word.

After demodulation and decoding stages, packets are received with some error probability called the packet error rate (PER). In general, the PER depends on the (post-detection) SNR, the packet length and the MCS. Assuming that the SNR is constant for the whole packet transmission, we use $PER^{(m)}(\gamma)$ to denote the PER when the SNR is γ and the MCS is m. Similarly, we use $PSR^{(m)}(\gamma) = 1 - PER^{(m)}(\gamma)$ to denote the packet success rate (PSR).

Figure 2.3 and Figure 2.4 depict the simulated PER for BPSK and 64-QAM modulation schemes and a turbo code with different

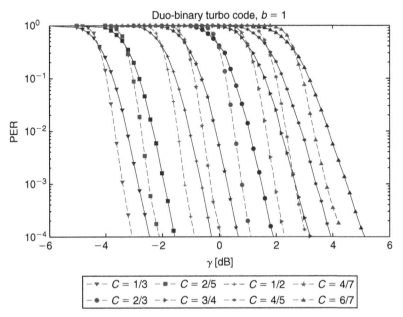

Figure 2.3 *PER curves for BPSK. Continuous lines: k = 288, dashed lines: k = 1152.*

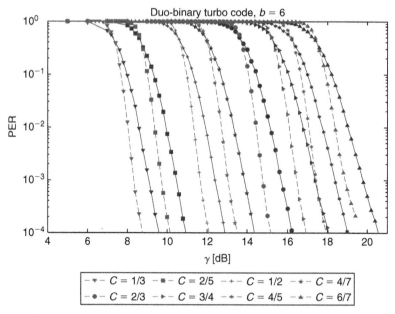

Figure 2.4 *PER curves for 64-QAM. Continuous lines: $k = 288$, dashed lines: $k = 1152$.*

code rates and packet lengths equal to 288 and 1152 information bits. That is, $k = 288$ and $k = 1152$ information bits. The applied channel code is a duo-binary turbo code [4] with another code rate 1/3 according to [5] and rate-compatible puncturing.

2.5.1   Analytical approximation of PER curves

For system-level simulations as well as for mathematical analysis of coded transmission systems, it is often of great benefit to have closed-form equations for the packet error probability. Since a direct mathematical derivation is too complex for any realistic coding scheme, we focus on an approximation based on the simulation results which are partially given in Figure 2.3 and Figure 2.4.

The packet error probability is approximated by an exponential function with two parameters, i.e.

$$PER^{(m)}(\gamma) \simeq \begin{cases} 1 & \text{for } \gamma \leq \gamma^{(m)} \\ \exp(-\alpha^{(m)}(\gamma - \gamma^{(m)})) & \text{for } \gamma \geq \gamma^{(m)} \end{cases} \quad (2.6)$$

A similar exponential approximation appears in [6]. However, the approximation in [6] is based on the estimation of three different

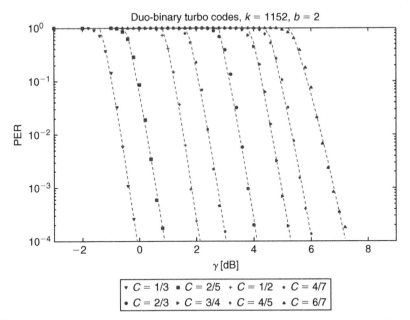

Figure 2.5 *Approximated PER curves (dotted lines) in comparison with simulation results (blue markers).*

parameters and can be simplified to (2.6) where only two parameters $\alpha^{(m)}$ and $\gamma^{(m)}$ are to be computed. If we denote the simulated SNRs and corresponding PERs by $\gamma_1, \gamma_2, \ldots, \gamma_n$ and $p_1, p_2, \ldots, p_n$, the two parameters $\alpha^{(m)}$ and $\gamma^{(m)}$ are calculated such that the relative quadratic error

$$\sum_{i=1}^{n} \frac{\left(p_i - PER^{(m)}(\gamma_i)\right)^2}{p_i^2} \qquad (2.7)$$

is minimized. Note that for $\alpha^{(m)} \to \infty$ the PER approximation becomes the extensively used step function approximation.

Figure 2.5 shows that the approximation fits very well to the simulation results. Only for very high PER is the approximation rather coarse. The parameters for the approximation of simulation results are given in Table 2.1.

2.6 The throughput

Shannon's capacity formula assumes that an optimal error-free MCS is used. However, as explained in previous sections, transmission

Table 2.1   *Parameters $\alpha^{(m)}$ and $\gamma^{(m)}$ for approximation of PER curves with expression.*

k	$b^{(m)}$	$c^{(m)}$	$\alpha^{(m)}$	$10\log_{10}(\gamma^{(m)})$	k	$b^{(m)}$	$c^{(m)}$	$\alpha^{(m)}$	$10\log_{10}(\gamma^{(m)})$
288	2	1/3	21.1525	−1.47 dB	1152	2	1/3	37.3454	−1.34 dB
	2	2/5	17.7710	−0.50 dB		2	2/5	28.6631	−0.46 dB
	2	1/2	13.0877	0.75 dB		2	1/2	22.7759	0.87 dB
	2	4/7	10.5338	1.68 dB		2	4/7	17.4025	1.71 dB
	2	2/3	7.3723	2.73 dB		2	2/3	13.5356	2.79 dB
	2	3/4	5.0264	3.71 dB		2	3/4	9.8805	3.89 dB
	2	4/5	4.0337	4.30 dB		2	4/5	7.8614	4.52 dB
	2	6/7	2.8135	5.07 dB		2	6/7	4.9148	5.31 dB
	4	1/3	6.0820	3.31 dB		4	1/3	10.7319	3.41 dB
	4	2/5	4.8686	4.46 dB		4	2/5	8.4153	4.53 dB
	4	1/2	3.2201	5.96 dB		4	1/2	5.8436	6.11 dB
	4	4/7	2.4331	7.12 dB		4	4/7	4.1306	7.16 dB
	4	2/3	1.6545	8.50 dB		4	2/3	3.2371	8.63 dB
	4	3/4	1.0728	9.69 dB		4	3/4	2.1561	9.90 dB
	4	4/5	0.8642	10.46 dB		4	4/5	1.6916	10.68 dB
	4	6/7	0.5777	11.30 dB		4	6/7	0.9705	11.54 dB
	6	1/3	2.2671	7.08 dB		6	1/3	4.1137	7.15 dB
	6	2/5	1.5902	8.32 dB		6	2/5	2.9282	8.50 dB
	6	1/2	1.0273	10.32 dB		6	1/2	1.8349	10.43 dB
	6	4/7	0.7162	11.69 dB		6	4/7	1.2743	11.75 dB
	6	2/3	0.4407	13.35 dB		6	2/3	0.9115	13.53 dB
	6	3/4	0.2815	14.90 dB		6	3/4	0.5582	15.09 dB
	6	4/5	0.2223	15.85 dB		6	4/5	0.4321	16.04 dB
	6	6/7	0.1448	16.86 dB		6	6/7	0.2554	17.09 dB
	8	1/3	0.8506	10.16 dB		8	1/3	1.5974	10.32 dB
	8	2/5	0.6093	12.03 dB		8	2/5	1.1152	12.09 dB
	8	1/2	0.3410	14.35 dB		8	1/2	0.6524	14.51 dB
	8	4/7	0.2203	15.94 dB		8	4/7	0.4343	16.13 dB
	8	2/3	0.1301	18.04 dB		8	2/3	0.2754	18.22 dB
	8	3/4	0.0742	19.76 dB		8	3/4	0.1577	20.08 dB
	8	4/5	0.0569	20.90 dB		8	4/5	0.1109	21.15 dB
	8	6/7	0.0361	22.09 dB		8	6/7	0.0652	22.44 dB

errors might occur in practical systems. Let us define the function $\mathbf{1}\{Z_i = 1\}$ as the indicator function of the variable Z_i that takes value 1 every time a packet is correctly received, that is, the variable Z_i is a Bernoulli distributed random variable with

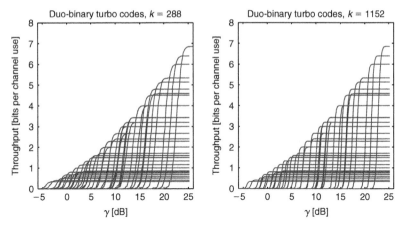

Figure 2.6 *Throughput of many MCS.*

probability $PSR^{(m)}(\gamma)$. Assume that a packet is transmitted at time instant t. Then, the amount of error-free data that the MAC/Link layer delivers to the upper layers of the communication system at time instant t is

$$\eta^{(m)}(\gamma) = R^{(m)}\mathbf{1}\{Z_i = 1\} \qquad (2.8)$$

where m is the MCS and $R^{(m)}$ is the effective transmission rate (2.5) for that MCS. Recall that (2.8) is expressed in bits per channel use.

One of the differences between (2.2) and (2.8) is that (2.8) is an instantaneous value that varies in time, even when channel conditions, given by γ, are constant. Therefore, throughput is commonly defined as the expectation of (2.8), which is given by

$$\overline{\eta}^{(m)}(\gamma) = E_t\left\{\eta^{(m)}(\gamma)\right\} = R^{(m)}PSR^{(m)}(\gamma) \qquad (2.9)$$

The PER curves, which are printed in Figure 2.3 and Figure 2.4 as a function of the SNR, can also be plotted as throughput over SNR, where throughput is defined as (2.9). This is plotted in Figure 2.6, where the throughput of many MCSs is represented. It can be seen immediately that many MCSs can be excluded since they require higher SNR for achieving the same or lower throughput than others. Moreover, we observe that for each value of SNR p0290 there is an optimal MCS that maximizes throughput. This motivates the idea of link adaptation.

s0080
2.7 Link adaptation

p0300 It is well known that, compared to the wireline alternatives, wireless links are severely degraded due to channel variability. Link adaptation refers to the capability of the system to adapt parameters such as modulation and coding to the quality of the wireless link. Although conventional radio communications systems were designed to adapt to the average channel quality, the great advantages of link adaptation come with the availability of modern systems to be able to follow the rapid variations of the channels.

p0310 Adapting the MCS, to face off the wireless channel variability has been widely studied [7–11]. Note that some of these works also consider transmitted power as an additional parameter to be adapted. The basic premise of link adaptation is to compensate channel quality degradation by adjusting the rate and the reliability of the transmission. Hence, when the channel experiences a deep fade the most reliable coding rates and data modulations are used and, conversely, when channel experiences a peak the MCS that offers the highest transmission rate is chosen.

p0320 We measure the quality of the link through the SNR, or in case of interferences, through the SNIR. Then, given a value of SNR, the MCS is adapted in order to optimize a given utility function, e.g. throughput, BER, PER or functions of them. Many link adaptation studies have been carried out from a PHY layer perspective considering BER as the utility function to be optimized, but recently some work has been also done on studying the impact of link adaptation at the MAC/Link layer. The works [12–15] consider link adaptation strategies based on throughput maximization instead of BER requirements. For instance, this is also considered by the high speed downlink packet access (HSDPA) proposal [16].

p0330 Clearly, link adaptation requires that channel state information (CSI) is available at the transmitter side. In time division duplexing (TDD) systems, it is sometimes assumed that the channel from the transmitter to the receiver is very similar to the channel from the receiver to the transmitter. Alternatively, a feedback channel can be designed so the channel is estimated at the receiver and fed back to the transmitter. How to design a feedback channel and its effects on system performance have been widely studied but is beyond the scope of this book.

Assume an ideal feedback channel so that the value of the SNR is perfectly known at the transmitter side. Then, throughput with link adaptation can be formulated as

$$\bar{\eta}(\gamma) = R^{(m^*)} PSR^{(m^*)}(\gamma) \tag{2.10}$$

where m^* is the optimal MCS given by

$$m^* = \arg\max_m R^{(m)} PSR^{(m)}(\gamma) \tag{2.11}$$

Assume that MCS schemes are ordered such that $R^{(m-1)} < R^{(m)} < R^{(m+1)}$. Therefore, because the PSR is a decreasing function that ranges between 1 and 0, there exists an SNR threshold $\gamma_{th}^{(m)}$ such that

$$R^{(m-1)} PSR^{(m-1)}(\gamma_{th}^{(m)}) = R^{(m)} PSR^{(m)}(\gamma_{th}^{(m)}) \tag{2.12}$$

and according to (2.11), an MCS m is optimal in the SNR range

$$\gamma \in \left[\gamma_{th}^{(m)}, \gamma_{th}^{(m+1)}\right) \tag{2.13}$$

which can be computed offline. An example is shown in Table 2.2, where the number of MCSs is limited to eight with respect to those in Figure 2.6 and where we assumed $R^{(0)} = 0$. The SNR level that

Table 2.2 *SNR thresholds for throughput maximization.*

m	1	2	3	4	5	6	7	8
MCS($b^{(m)}$, $c^{(m)}$)	$\left(1,\frac{1}{3}\right)$	$\left(2,\frac{1}{2}\right)$	$\left(2,\frac{3}{4}\right)$	$\left(4,\frac{1}{2}\right)$	$\left(4,\frac{3}{4}\right)$	$\left(6,\frac{2}{3}\right)$	$\left(6,\frac{6}{7}\right)$	$\left(8,\frac{4}{5}\right)$
Rate $R^{(m)}$	0.33	1	1.5	2	3	4	5.14	6.4
$k = 288, \gamma_{th}^{(m)}$		0.7 dB	4.2 dB	6.6 dB	10.3 dB	14.1 dB	17.9 dB	22 dB
$k = 1152, \gamma_{th}^{(m)}$		0.9 dB	4.1 dB	6.5 dB	10.2 dB	13.9 dB	17.7 dB	21.8 dB

causes a change in the optimal MCS corresponds to the SNR level where throughput curves in Figure 2.6 cross, that is, $\gamma_{th}^{(m)}$ in (2.12).

s0090

2.7.1 Analytical approximation of throughput envelope

p0370

Throughput expression given by (2.10) corresponds to the envelope in Figure 2.6. Analytically, the envelope curve can be approximated by a modified capacity formula of the kind

$$\overline{\eta}(\gamma) = \alpha_1 \log_2(1 + \alpha_2 \gamma) \qquad (2.14)$$

p0380

Parameters α_1 and α_2 are obtained through simulations and depend, among others, on the number of MCSs. An example of eigth MCSs is plotted in Figure 2.7. Of course, the fewer the number of MCSs, the less signaling overhead. However, this is at the cost of increased granularity.

p0390

Furthermore, there is also a gap between the information theory limit and throughput and we can clearly observe that the relationship between SNR and capacity is not the same as that between SNR and throughput. As a consequence, an increase or decrease of the SNR does not increase or decrease capacity and throughput in

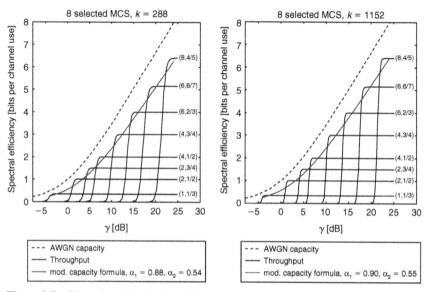

f0070 **Figure 2.7** *Throughput for 8 selected MCS.*

the same way. This means that, from a system point of view, decisions on resource allocation should differ depending on whether we consider capacity or throughput as figures of merit.

2.7.2 Quality of service requirements

Most multimedia applications such as VoIP, video streaming and conferencing require that a certain level of performance is guaranteed. Such level of performance is known as QoS. At the PHY and MAC/Link layers, QoS requirements are typically translated into minimum throughput, maximum delay and minimum SNR or PSR requirements.

Although current wireless communication standards provide architectures and signaling to perform resource allocation with QoS constraints, the ultimate resource allocation mechanisms and algorithms that provide QoS are left to the manufacturer's implementation. However, how to perform resource allocation with QoS requirements in wireless systems is not an easy task [23, 24] due to the classical wireless channel impairments associated with its fading and broadcast nature. Consequently, the general framework to evaluate resource allocation in wireless channels that will be presented along with this book is of special interest.

In this chapter the QoS metric analyzed is throughput whereas other QoS metrics such as delay will be considered later. Nevertheless, maximizing throughput might not be enough for some practical applications but minimum QoS requirements in terms of a minimum PSR threshold PSR_{QoS} might be also required. Recall that our aim is to maximize throughput while other objectives such as minimize the total transmitted power could be also foreseen [25].

Accounting for link adaptation and QoS requirements on the PSR, throughput (2.10) is reformulated as

$$\overline{\eta}(\gamma) = R^{(m^*)} PSR^{(m^*)}(\gamma) \tag{2.15}$$

where

$$m^* = \arg \max_{m \atop s.t.\ PSR^{(m)}(\gamma) \geq PSR_{QoS}} R^{(m)} PSR^{(m)}(\gamma) \tag{2.16}$$

p0440   Note that according to (2.6), the QoS requirement PSR_{QoS} corresponds to an SNR level $\gamma_{QoS}^{(m)}$ given by the inverse function of (2.6). Therefore, expression (2.16) can be rewritten as

$$m^* = \arg \max_{\substack{m \\ s.t.\ \gamma \geq \gamma_{QoS}^{(m)}}} R^{(m)} PSR^{(m)}(\gamma) \tag{2.17}$$

p0450   Similarly to (2.13), the SNR range where the MCS m is optimal is given by

$$\gamma \in \left[\gamma_{QoS}^{(m)}, \gamma_{QoS}^{(m+1)}\right) \tag{2.18}$$

p0460   In best effort conditions, i.e. when there are no QoS requirements on the PSR, it is clear that $\gamma_{QoS}^{(m)} = \gamma_{th}^{(m)}$. An example of the optimal values m^* is shown in Table 2.3 for $PSR_{QoS} = 0.999$. Note that a QoS requirement of $PSR_{QoS} = 0.999$ implies that a maximum PER of 10^{-3} is accepted by the application level.

p0470   An example of throughput with and without QoS requirements is depicted in Figure 2.8. Note that according to that figure, the step function approximation of the PSR seems more accurate as the QoS requirements in terms of minimum PSR increase. However, also note that the difference between considering QoS requirements or not when maximizing throughput is only on the SNR range where a given MCS is optimal but not on the formulation

t0030   **Table 2.3** *SNR thresholds for 8 selected MCS and $PSR_{QoS} = 0.999$.*

M	1	2	3	4	5	6	7	8
MCS $(b^{(m)}, c^{(m)})$	$\left(1, \frac{1}{3}\right)$	$\left(2, \frac{1}{2}\right)$	$\left(2, \frac{3}{4}\right)$	$\left(4, \frac{1}{2}\right)$	$\left(4, \frac{3}{4}\right)$	$\left(6, \frac{2}{3}\right)$	$\left(6, \frac{6}{7}\right)$	$\left(8, \frac{4}{5}\right)$
Rate $R^{(m)}$	0.33	1	1.5	2	3	4	5.14	6.4
$k = 288$, $\gamma_{QoS}^{(m)}$	$-2.8\,$dB	$2.4\,$dB	$5.8\,$dB	$8.0\,$dB	$12.0\,$dB	$15.8\,$dB	$19.8\,$dB	$24.0\,$dB
$k = 1152$, $\gamma_{QoS}^{(m)}$	$-3.2\,$dB	$1.8\,$dB	$5.0\,$dB	$7.2\,$dB	$11.2\,$dB	$14.8\,$dB	$19.0\,$dB	$22.8\,$dB

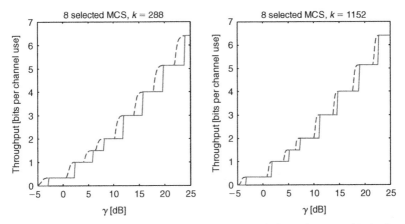

Figure 2.8 *Throughput with 8 selected MCS without QoS requirements (dashed) and with QoS requirement of $PSR_{QoS} = 0.999$ (solid).*

of the problem itself. Consequently, in the following chapters no additional formulation will be required in order to include QoS requirements in the analysis.

2.8 The average spectral efficiency

The wireless channel is a time-varying fading channel such that the spectral efficiency metrics capacity and throughput change in time according to a random fading process. Assuming that the channel fading process is time varying and ergodic, the time average capacity is also the ergodic capacity and is given by averaging (2.2) either over time or over the fading statistics

$$\overline{C} = E_{\gamma}\{\log_2(1 + \gamma)\} \tag{2.19}$$

The ergodic capacity (2.19) can be achieved without instantaneous feedback of the channel but with a capacity-achieving codeword that is long enough so that the transmitted data is encoded over all the possible channel fading states. Although this can be easily assumed in rapid varying (or fast fading) channels, this is not necessarily true in slow fading channels. Nevertheless, short codes could be used in slow fading channels and, hence, delay could be reduced at the expense of experiencing transmission errors. Errors occur when the SNR is below a threshold determined by the code. In that case, we say that the system is in outage. Besides, the probability of the SNR being below the threshold is the outage probability. Clearly, the shorter the code, the lower the delay but the higher the outage probability [22].

p0500   Similarly, by channel ergodicity, the average throughput is obtained by averaging (2.10) over the channel fluctuations

$$\bar{\eta} = E_\gamma \left\{ R^{(m^*)} PSR^{(m^*)}(\gamma) \right\} \qquad (2.20)$$

where the optimal MCS is given by (2.11) or (2.17) depending on the QoS constraints.

p0510   In the literature, the most commonly used model for the wireless channel fading process is the Rayleigh fading which is most applicable when there is no dominant propagation along a line of sight between the transmitter and receiver. If we define the average SNR as the ratio between the signal power P and the noise power σ^2 in the AWGN channel, expressed as $\bar{\gamma} = P/\sigma^2$, the cumulative density function (c.d.f.) of the SNR in a Rayleigh fading channel is given by $F(\gamma) = 1 - e^{-\gamma/\bar{\gamma}}$.

p0520   The average throughput in a Rayleigh fading channel is given in Figure 2.9. The average capacity is also plotted for comparison.

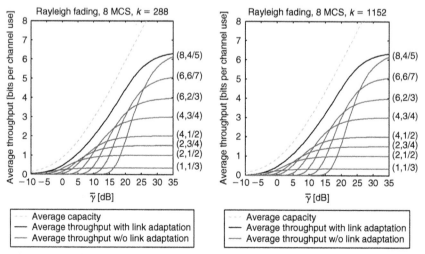

f0090   **Figure 2.9** *Average throughput in Rayleigh fading channel with link adaptation (8 MCS) and without link adaptation.*

For each MCS, the average throughput without link adaptation corresponds to the average of the throughput plotted in Figure 2.7. The average throughput with link adaptation corresponds to the average of the envelope in Figure 2.7 or, equivalently, the curve in Figure 2.8 for no QoS constraints. Clearly, the fact of choosing the best transmission mode every time instant, i.e. link adaptation, offers a significant gain on the average throughput.

2.9 Summary

In this chapter the concept of spectral efficiency was introduced showing the differences between taking a totally PHY layer point of view and a PHY–MAC cross-layer point of view. In particular, we introduced the necessary formulation for the analysis of system performance in terms of throughput and showed the differences between capacity and throughput. Taking into account the variability of the wireless channel, the strategy of link adaptation was also presented and QoS requirements were also considered in this chapter. Furthermore, the concept of average spectral efficiency was also presented.

This chapter was devoted to analyzing a single link, that is, a source and a destination. However, wireless communications systems typically involve the communication of multiple users and then parameters such as interferences come into play. An efficient way of allocating resources such as the transmitted power might help to mitigate interferences and to increase system performance. Therefore, these are the topics tackled in the following chapters of this book.

References

[1] C.E. Shannon, A mathematical theory of communication, *Bell System Technical Journal*, Vol. 27, pp. 379–423 and 623–656, July and Oct. 1948.

[2] J.G. Proakis, *Digital Communications*, 3rd ed., McGraw-Hill, New York, 1995.

[3] IEEE Std 802.11a, Supplement to part 11: Wireless LAN, Medium Access Control (MAC) and Physical Layer (PHY) Specifications: High-Speed Physical Layer in the 5 GHz Band, IEEE Std 802.11a-1999, Sept. 1999.

[4]   C. Berrou, et al., Duo-binary turbo codes associated with high-order modulation, *ESA TTC '01*, Oct. 2001.

[5]   ETSI EN 301 790 V1.4.1 (2005–09), Digital Video Broadcasting (DVB); interaction channel for satellite distribution systems, Sept. 2005.

[6]   Q. Liu, S. Zhou and G.B. Giannakis, Cross-layer combining of adaptive modulation and coding with truncated ARQ over wireless links, *IEEE Trans. Wireless Comm.*, Vol. 3, Issue 5, pp. 1746–1755, Sept. 2004.

[7]   T. Ue, S. Sampei and N. Morinaga, Symbol rate and modulation level controlled adaptive modulation/TDMA/TDD for personal communication systems, *Proc. IEEE Vehicular Technology and Communications (VTC)*, July 1995.

[8]   A.J. Goldsmith and S. Chua, Adaptive coded modulation for fading channels, *IEEE Trans. Comm.*, Vol. 46, pp. 595–602, May 1998.

[9]   D.L. Goeckel, Adaptive coding for time-varying channels using outdated fading estimates, *IEEE Trans. Comm.*, Vol. 47, pp. 844–855, June 1999.

[10] C. Köse and D.L. Goeckel, On power adaptation in adaptive signaling systems and resource allocation in cellular and packet radio networks, *IEEE Trans. Comm.*, Vol. 48, pp. 1769–1773, Nov. 2000.

[11] S.T. Chung and A.J. Goldsmith, Degrees of freedom in adaptive modulation: a unified view, *IEEE Trans. Comm.*, Vol. 49, Issue 9, Sept. 2001.

[12] D. Qiao, S. Choi and K.G. Shin, Goodput analysis and link adaptation for IEEE 802.11a wireless LANs, *IEEE Trans. Mobile Computing*, Vol. 1, Issue 4, Dec. 2002.

[13] J. del Prado and S. Choi, Link adaptation strategy for IEEE 802.11 WLAN via received signal strength measurement, *ICC 2003*, Vol. 2, May 2003.

[14] W. Wang, et al, Evaluations of a 4G uplink system based on adaptive single-carrier TDMA, *Proc. Vehicular Technology and Communications (VTC)*, Sept. 2005.

[15] T. Yoo, R. Lavery, A. Goldsmith and D. Goodman, Throughput optimization using adaptive techniques, submitted to *Communication Letters*, 2006.

[16] 3GPP TS 25.848, 3rd Generation Partnership Project; Technical Specification Group Radio Access Network; physical layer aspects of UTRA high speed downlink packet access, Release 4, March 2001.

[17] T. Richardson and R. Urbanke, Modern coding theory, in preparation.

[18] T. Moon, *Error Correction Coding: Mathematical Methods and Algorithms*, John Wiley & Sons, 2005.

[19] D.J.C. MacKay and R.M. Neal, Near Shannon limit performance of low density parity-check codes, *Electronics Letters*, Vol. 32, Issue 18, 29 Aug. 1996.

[20] R.G. Gallager, *Low Density Parity-Check Codes*, MIT Press, Cambridge, MA, 1963.

[21] C. Berrou, A. Glavieux and P. Thitimajshima, Near Shannon limit error-correcting coding and decoding: turbo-codes, *IEEE ICC*, May 1993.

[22] A. Goldsmith, *Wireless Communications*. Cambridge University Press, 2005.

[23] Q. Liu, X. Wang and G.B. Giannakis, A cross-layer scheduling algorithm with QoS support in wireless networks, *IEEE Trans. Vehicular Technology*, Vol. 55, Issue 3, pp. 839–847, May 2006.

[24] Q. Liu, S. Zhou and G.B. Giannakis, Cross-layer modeling of adaptive wireless links for QoS support in heterogeneous wired-wireless networks, *ACM/Kluwer Journal on Wireless Networks (WINET)*, Vol. 12, pp. 427–437, May 2006.

[25] B. Zerlin, M.T. Ivrlac, J.A. Nossek and A. Klein, On cross-layer assisted resource allocation in HSDPA, *Proc. 12th IEEE International Conference on Telecommunication*, May 2005.

3

The cross-layer resource allocation problem

In multi-user communication systems, different users or terminals share the same radio channel to communicate among them. Then, in the context of multi-user systems, a major question of importance is how to allocate resources in an optimal fashion so that ultimate limits could be achieved under idealized assumptions and given constraints. The insights gained from these considerations can serve as a guideline for the design of real systems and development of signal processing algorithms.

To present a general framework for resource allocation is not an easy task, especially if a cross-layer perspective is desired by considering parameters such as channel coding, data modulation and link adaptation. Therefore, in this book we perform some simplifications on the general cross-layer resource allocation framework for MIMO systems by considering the uplink of multi-user SIMO wireless systems and the downlink of multi-user MISO ones. Clearly, any of these two cases can be further simplified to multi-user SISO systems.

The full characterization of the achievable system performance requires the identification of a *region*, defined as the set of achievable spectral efficiencies in Chapter 2. Such a region (capacity region or throughput region) reflects the trade-off among the individual spectral efficiencies of the different users competing for the limited resources. For the Gaussian K-user channel, it is well known that the capacity region is a convex region in the K-dimensional space.

p0040   Some specific spectral efficiency distributions on the boundary of the capacity or throughput region, corresponding to some desirable working points, should be identified. The mapping between the resource allocation and the obtained spectral efficiency distribution is dependent on the communication scenario. In every scenario, a specific constrained optimization problem can be obtained, and can be solved by appropriate methods, depending, among others, on the shape of the region (convexity, etc.).

s0010   ## 3.1 Allocating resources: time, power, space and frequency

p0050   In a multi-user wireless system, different users suffer from different and independent channel conditions resulting in, at a given time instant, some users experiencing better conditions than others. Users with better channel conditions are the ones that can perform a more efficient channel use. A scheduler with CSI is able to exploit this situation by scheduling the best user and thus realize a multi-user diversity (MUD) gain.

p0060   Traditionally, it was assumed that either transmitting data to simultaneous users or the simultaneous reception of multiple users was not possible due to the strong interference component of the received signal. However, modern signal processing techniques such as successive interference cancellation (SIC) allow the scheduler to select a set of multiple users for a simultaneous channel use. Moreover, by means of multiple antennas, independent channel conditions among different transmit and receive antenna pairs can be created. Therefore, multiple antenna transceivers with advanced signal processing introduce an additional domain, the spatial domain, for the scheduler to allocate among terminals.

p0070   By means of orthogonal frequency division multiplexing (OFDM) or multi-carrier modulation, modern wireless systems are likely to exploit resource allocation in the frequency domain in addition to temporal and spatial domains. In general, resource allocation in multi-user MIMO communications implies not only scheduling the transmission of several data streams over multiple antennas and frequency subchannels but also the allocation of the available transmission power in time and over all the antennas and frequency subchannels.

Typically, the main objective is to optimize the region of achievable spectral efficiencies and/or a system utility function of spectral efficiencies. However, in order to work with the different views of spectral efficiency as defined in Chapter 2, considering channel coding, data modulation and link adaptation, the post-processing SNIR is required rather than the pre-processing SNIR which further complicates the resource allocation process. With the purpose of presenting a general but simple formulation of the post-processing SNIR, we limit ourselves to studying the multi-user MISO and SIMO scenarios. Hence, our general scenario is composed of a base station (BS) or access point (AP) provided with M antennas and K single antenna user terminals. Moreover, we further assume that the number of frequency subchannels is equal to one and refer to some chapters at the end of the book to tackle the resource allocation problem in OFDM systems.

3.2 Signal model for multi-user SIMO multiple access channel

Let us consider a multi-user SIMO multiple access channel (or uplink) such as the one in Figure 3.1, where K single antenna terminals communicate with the AP provided with M antennas. In the uplink, we assume that the channel state information (CSI) is available at the AP and that the AP allocates resources via an ideal signaling channel in the downlink.

The received signal vector $\mathbf{r}$ is modeled as

$$\mathbf{r} = \mathbf{Hs} + \mathbf{w} \tag{3.1}$$

where the vector $\mathbf{s}$ is the transmitted symbol vector being s_k the transmitted symbol of user k and $\mathbf{H}$ is the $M \times k$ flat fading channel matrix with complex valued entries $h_{m,k}$ corresponding to the channel fading between the user's antenna and each of the antennas at the AP. Note that each column $\mathbf{h}_k$ of the channel matrix $\mathbf{H}$ is the kth user $M \times 1$ SIMO channel. The vector $\mathbf{w}$ is a complex valued, background Gaussian noise with zero mean and variance σ^2.

Resource allocation signalling via downlink   CSI feedback (or from uplink measurement in TDD)

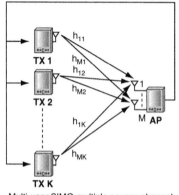 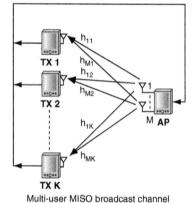

Multi-user SIMO multiple access channel   Multi-user MISO broadcast channel
(uplink)   (downlink)

Figure 3.1   *Multi-user SIMO multiple access (left) and MISO broadcast (right) channels.*

The received signal is processed at the AP by a $K \times M$ processing matrix $\mathbf{V}$ so that the received signal becomes

$$
\begin{aligned}
\mathbf{y} &= \mathbf{Vr} \\
&= \mathbf{VHs} + \mathbf{Vw} \\
&= \mathbf{VHs} + \mathbf{z}
\end{aligned}
\tag{3.2}
$$

Operating this expression we obtain that the kth component of the vector $\mathbf{y}$ is

$$
y_k = \mathbf{v}_k^t \mathbf{h}_k s_k + \sum_{\substack{k' \\ k' \neq k}} \mathbf{v}_k^t \mathbf{h}_{k'} s_{k'} + z_k
\tag{3.3}
$$

where $\mathbf{v}_k^t$ is the transpose of $\mathbf{v}_k$. From (3.3), the SNIR corresponding to the signal of user k can be expressed as

$$
\gamma_k = \frac{\left|\mathbf{v}_k^t \mathbf{h}_k\right|^2 p_k}{\sigma^2 + \sum_{\substack{k' \in \mathbb{K} \\ k' \neq k}} \left|\mathbf{v}_k^t \mathbf{h}_{k'}\right|^2 p_{k'}}
\tag{3.4}
$$

where $\mathbb{K}$ is the set of transmitting users defined by $\mathbb{K} = \{k \mid k \in \{1, \ldots, K\}; p_k > 0\}$. Note that the total transmitted power corresponding to user k is governed by $p_k = E\{s_k s_k^*\}$.

3.2.1 The successive interference cancellation (SIC) receiver

The signal expressed in (3.2) is specific for the case where all the components of **r** are processed and decoded simultaneously. This conventional receiver presents the near–far problem where a user close to the AP, i.e. with a strong channel, might be very harmful for a user far from the AP.

Rather than decoding every user treating the interference as noise, an alternative is the SIC receiver that iteratively decodes the components of the signal vector **s** by subtracting, at each iteration, the interference created by the already processed components. Therefore, by processing the strongest user first, the SIC receiver helps to overcome the near–far problem.

Define $\pi = \{\pi_1, ..., \pi_K\}$ as the SIC decoding order such that user k with $\pi_i = k$ is the ith user to be decoded. Then, the component of the signal vector $\mathbf{y}^{SIC}$ corresponding to the kth user is

$$
\begin{aligned}
y_k^{SIC} &= \mathbf{v}_k^t \mathbf{h}_k s_k + \sum_{\substack{k' \in \mathbb{K} \\ k' \neq k}} \mathbf{v}_k^t \mathbf{h}_{k'} s_{k'} - \sum_{\substack{k' \in \mathbb{K} \\ k' \in \{\pi_1, ..., \pi_{i-1}\}}} \mathbf{v}_k^t \mathbf{h}_{k'} \hat{s}_{k'}^{SIC} + z_k \\
&= \mathbf{v}_k^t \mathbf{h}_k s_k + \sum_{\substack{k' \in \mathbb{K} \\ k' \in \{\pi_{i+1}, ..., \pi_K\}}} \mathbf{v}_k^t \mathbf{h}_{k'} s_{k'} + z_k
\end{aligned}
\tag{3.5}
$$

where $\hat{s}_{k'}^{SIC}$ is the estimated transmitted symbol given $y_{k'}^{SIC}$. Note that the noise component z in (3.5) is different to that in (3.3). However, noise is always assumed to be Gaussian. Therefore, it is easy to observe that the SNIR expression (3.4) also applies in this if $k' \neq k$ is substituted by $k' \in \{\pi_{i+1}, ..., \pi_K\}$.

Unfortunately, the SIC receiver presents some important issues that make its practical implementation very difficult [1]. Probably the more important are the error propagation and the effect of imperfect channels estimates. Note that if a user is decoded with an error, the estimated symbol $\hat{s}_{k'}^{SIC}$ is different to the real transmitted symbol s_k, and then all the users decoded later are very likely to be decoded incorrectly. Clearly, one source of incorrect decoding is the use of an imperfect channel estimate.

s0040

3.3   Signal model for multi-user MISO broadcast channel

p0180

In the multi-user MISO broadcast channel (or downlink) as the one in Figure 3.1, an AP provided with M antennas communicates with K single antenna terminals. A perfect instantaneous feedback channel is assumed for enhancing the communication. By means of the messages on the feedback channel, the AP acquires the users' CSI required for implementing the resource allocation.

p0190

In the multi-user MISO broadcast channel the signal processing is performed at the AP and each user receives its signal; users cannot cooperate in the detection process. Each symbol s_k of the transmitted signal vector $\mathbf{s}$ is processed by a vector $\mathbf{v}_k$ and transmitted through the MISO channel. Considering the same channel matrix $\mathbf{H}$ as in the multiple access channel, the received signal at user k is given by

$$y_k = \mathbf{v}_k^t \mathbf{h}_k s_k + \sum_{\substack{k' \in \mathbb{K} \\ k' \neq k}} \mathbf{v}_{k'}^t \mathbf{h}_k s_{k'} + z_k \tag{3.6}$$

p0200

The main difference between the multiple access channel signal model (3.3) and the broadcast signal model (3.6) is that in the broadcast channel, the signal of interest s_k and the interferences $s_{k'}$ travel through the same channel $\mathbf{h}_k$. The duality between the multiple access channel and the broadcast channel has been studied in the literature [1].

p0210

By imposing that $\mathbf{V}$ is normalized so that $\mathbf{V}\mathbf{V}^H = \mathbf{I}$, the total transmitted power corresponding to user k is governed by $p_k = E\left\{s_k s_k^*\right\}$. Then, the signal to noise and interference ratios corresponding to the signal of user k can be expressed as

$$\gamma_k = \frac{\left|\mathbf{v}_k^t \mathbf{h}_k\right|^2 p_k}{\sigma^2 + \sum_{\substack{k' \in \mathbb{K} \\ k' \neq k}} \left|\mathbf{v}_{k'}^t \mathbf{h}_k\right|^2 p_{k'}} \tag{3.7}$$

p0220

Following the same philosophy as for the SIC receiver in the multiple access channel, a sequential encoding could be used at the transmitter side so that before transmitting a symbol to a user this symbol is encoded using the symbols to be sent to other users. This process is known as dirty paper coding (DPC) [3] when multiple antennas are used. Then, an encoding order $\pi = \{\pi_1, \ldots, \pi_K\}$ determines that a user k with $\pi_i = k$ is the ith user to be encoded using information of users $\{\pi_1, \ldots, \pi_{i-1}\}$. In that way, the first encoded user suffers from interferences from all the other users and, as users are sequentially encoded, the interference of the previous encoded users can be eliminated at each iteration. The last encoded user is the one that does not suffer from any interference. The received vector components are given by

$$
\begin{aligned}
y_k^{DPC} &= \mathbf{v}_k^t \mathbf{h}_k s_k + \sum_{\substack{k' \in \mathbb{K} \\ k' \neq k}} \mathbf{v}_{k'}^t \mathbf{h}_k s_{k'} - \sum_{\substack{k' \in \mathbb{K} \\ k' \in \{\pi_1, \ldots, \pi_{i-1}\}}} \mathbf{v}_{k'}^t \mathbf{h}_k \hat{s}_{k'}^{DPC} + z_k \\
&= \mathbf{v}_k^t \mathbf{h}_k s_k + \sum_{\substack{k' \in \mathbb{K} \\ k' \in \{\pi_{i+1}, \ldots, \pi_K\}}} \mathbf{v}_{k'}^t \mathbf{h}_k s_{k'} + z_k
\end{aligned}
\tag{3.8}
$$

and similarly for the multiple access channel case, the SNIR expression (3.7) also applies in this case substituting $k' \neq k$ by $k' \in \{\pi_{i+1}, \ldots, \pi_K\}$.

0230 Either at the transmitter in the downlink MISO system or at the receiver in the uplink SIMO case, the signal of user k is processed by the signal processing vector $\mathbf{v}_k$ that represents the weight of the signal at each of the AP antennas. In Chapter 2, the SNR used to define the concept of spectral efficiency is the same as (3.4) and (3.7) but without interferences. Therefore, the theory in Chapter 2 also applies considering (3.4) and (3.7). Some examples on how to model interferences through different signal processing techniques are given in the annex to this chapter.

0240 However, in this chapter, rather than evaluating the spectral efficiency of a single user communication link, we evaluate a multiuser communication link and, hence, evaluate a set of spectral efficiencies. To deal with the set of possible spectral efficiencies is difficult because each user spectral efficiency depends on the interferences created among users which, in turn, are strongly dependent on the different resource allocation policies. Therefore, the system performance must be evaluated in terms of a region composed of all the sets of spectral efficiencies achievable by all the possible resource allocation policies. Once this region is obtained, the resource allocation problem is to find the optimal resource allocation policy that makes the system work at any given point of such a region. Clearly, before studying the region of spectral efficiencies and analyzing the resource allocation problem, the concept of resource allocation policy must be defined.

3.4 The resource allocation policy definition

Whatever is the case, uplink or downlink, the SNIR of one user is shown to be strongly dependent on the channel matrix $\mathbf{H}$ and on the power p_k allocated to the different users. Then, assuming a block fading channel such that a channel realization $\mathbf{H}$ is constant for a block duration of T seconds but changes independently every T seconds, we define a resource allocation policy as a power allocation procedure that every T seconds maps a new channel realization $\mathbf{H}$ to a power allocation vector $\mathbf{p(H)} = \{p_1, ..., p_k\}$.

Different power restrictions must be taken into account depending on whether we consider resource allocation in the uplink or in the downlink of the system. In particular, power restrictions in the downlink usually refer to total power restrictions at the AP with independence on the users that are served. That is

$$\sum_{k=1}^{K} p_k \leq P_{total} \tag{3.9}$$

where P_{total} is the total power budget. On the other hand, in the uplink, limitations are on the user terminal and, hence, power restrictions are individual. That is,

$$p_k \leq P \quad \forall\, k \in \{1, ..., K\} \tag{3.10}$$

where the power restriction P can be different among users. Moreover, average power constraints could be also considered leading to the conditions

$$E\left\{\sum_{k=1}^{K} p_k\right\} \leq P_{total} \tag{3.11}$$

and

$$E\{p_k\} \le P \quad \forall k \in \{1, \ldots, K\} \tag{3.12}$$

for downlink and uplink, respectively.

A resource allocation policy is completely characterized by the vector set $\mathbf{P} = \{\mathbf{p}(\mathbf{H}) : \mathbf{H} \in C^{M \times K}\}$ that contains all the power allocation vectors $\mathbf{p}(\mathbf{H})$ for all possible channel realizations $\mathbf{H}$. With restrictions (3.9) and (3.10) the vector set $\mathbf{P}$ is a convex set.

3.5 The spectral efficiency region and the optimal resource allocation policy

The concept of spectral efficiency described in Chapter 2, can be presented as a function of the user's post-processing SNIR. As shown throughout this chapter, the SNIR depends on both the channel realization $\mathbf{H}$ and the power allocation vector $\mathbf{p}(\mathbf{H})$. Therefore, we use $R_k(\mathbf{H}, \mathbf{p}(\mathbf{H}))$ to denote either information theory rate or throughput of the kth user when the SNIR is as (3.4) and (3.7). That is $R_k(\mathbf{H}, \mathbf{p}(\mathbf{H})) = \log_2(1 + \gamma_k)$ or $R_k(\mathbf{H}, \mathbf{p}(\mathbf{H})) = R^{(m^*)}$ $PSR^{(m^*)}(\gamma_k)$, respectively. Recall that m^* is the optimal modulation and coding scheme given the SNIR γ_k and $R^{(m^*)}$ is the effective transmission rate in bits per channel use corresponding to such modulation and coding scheme.

Our interest is twofold: first, we want to investigate what are the individual QoS constraints in terms of average spectral efficiency that can be guaranteed by any resource allocation policy $\mathbf{P}$. And second, we want to find the optimal resource allocation policy that makes our system work at any achievable point. We define the average spectral efficiency region as the convex hull of all the individual average spectral efficiencies that can be achieved by any resource allocation policy. Given a resource allocation policy $\mathbf{P}$, denote the average spectral efficiency vector by

$$\begin{aligned} \bar{\mathbf{R}} &= \left\{ \bar{R_1}, \ldots, \bar{R_K} \right\} \\ &= \left\{ E\{R_1(\mathbf{H}, \mathbf{p}(\mathbf{H}))\}, \ldots, E\{R_K(\mathbf{H}, \mathbf{p}(\mathbf{H}))\} \right\} \end{aligned} \tag{3.13}$$

p0300   Then, we can define the set of achievable average spectral efficiency vectors over all the resource allocation policies. This is formally expressed as

$$S = \bigcup_{\mathbf{P}\in\mathbb{P}} \left\{ \bar{\mathbf{R}} : \bar{R}_k = E\{R_k(\mathbf{H}, \mathbf{p}(\mathbf{H}))\}, k \in \{1, \dots, K\} \right\} \quad (3.14)$$

where **P** is the set of all possible resource allocation policies **P** with the corresponding power constraints. Recall that when either DPC or SIC strategies are used, the rate of a given user depends not only on the allocated power but also on the encoding/decoding order. Therefore, in that case, the region S is formulated as

$$S = \bigcup_{\mathbf{P}\in\mathbb{P}} \bigcup_{\pi\in\Pi} \left\{ \bar{\mathbf{R}} : \bar{R}_k = E\{R_k(\mathbf{H}, \mathbf{p}(\mathbf{H}))\}, k \in \{1, \dots, K\} \right\} \quad (3.15)$$

where Π is the set of all possible permutations on $\{1, \dots, K\}$ corresponding to the different encoding/decoding orders of the users.

p0310   The convex hull of the points in S over all possible independent input powers and encoding/decoding orders defines the average spectral efficiency region Ω. This is expressed as

$$\Omega = \left\{ \sum_{k=1}^{K} \theta_k \bar{R}_k \,\middle|\, 0 \le \theta_k \le 1, \sum_{k=1}^{K} \theta_k = 1, \right. \\ \left. \bar{R}_k \in S, k \in \{1, \dots, K\} \right\} \quad (3.16)$$

p0320   Note that the convex hull operation means that we not only include points in S, but also all their convex combinations. This is not surprising since the convex combinations can be achieved by time sharing.

p0330   The average spectral efficiency region Ω is convex with respect to θ_k and $\bar{R}_k$, and the boundary of the average spectral efficiency region can be traced out by means of a set of relative priority coefficients θ_k with $\sum_{k=1}^{K} \theta_k = 1$. Therefore, each point at the boundary of Ω either maximizes the linear combination of the user rates $\bar{R}_\theta = \sum_{k=1}^{K} \theta_k \bar{R}_k$ or is the result of time sharing between points that maximize such linear combination.

p0340   Thanks to the property of linearity it is true that $\sum_{k=1}^{K} \theta_k \bar{R}_k = E\left\{\sum_{k=1}^{K} \theta_k R_k\right\}$ and then the maximum of the average weighted sum rate, $\bar{R}_\theta = \sum_{k=1}^K \theta_k \bar{R}_k$, is obtained by maximizing the instantaneous weighted sum rate $R_\theta(\mathbf{H}, \mathbf{p}(\mathbf{H})) = \sum_{k=1}^K \theta_k R_k(\mathbf{H}, \mathbf{p}(\mathbf{H}))$ at each channel realization $\mathbf{H}$. Therefore, any point at the boundary of Ω is achieved by a vector of relative priorities $\theta = \{\theta_1, \ldots, \theta_K\}$ and a resource allocation policy described by $\mathbf{p}_\theta^*(\mathbf{H})$ is given by

$$\mathbf{p}_\theta^*(\mathbf{H}) = \arg \max_{\mathbf{p}(\mathbf{H})} \left(\max_\pi \sum_{k=1}^K \theta_k R_k(\mathbf{H}, \mathbf{p}(\mathbf{H})) \right) \qquad (3.17)$$

and subject to the corresponding power restrictions.

In consequence, the solution to (3.17) indicates the power allocation that must be performed on every channel realization so the system works at a desired point of the average spectral efficiency region. It is worth recalling that some points are directly achieved by a given priority vector and allocating power according to the solution to (3.17) and other points require for a time sharing between different power allocation policies corresponding to different priority vectors.

Although, the main discussion in the following chapters will be in the form of the optimal resource allocation policy $\mathbf{p}_\theta^*(\mathbf{H})$ in different multi-user wireless communication scenarios, the most important idea in this chapter is to show that by governing the vector of relative priorities θ we could ideally move through all the points in the average throughput region. For instance, by setting all users with equal relative priorities, $\mathbf{p}_\theta^*(\mathbf{H})$, would give us the optimal resource allocation that maximizes the sum of average spectral efficiencies. In order to simplify the analysis, note that $\max \sum_{k=1}^K \theta_k R_k(\mathbf{H}, \mathbf{p}(\mathbf{H})) = \max \sum_{k=1}^K L\theta_k R_k(\mathbf{H}, \mathbf{p}(\mathbf{H}))$ where L is any positive arbitrary scaling factor. Therefore, restrictions $\theta_k \leq 1$ and $\sum_{k=1}^K \theta_k = 1$ can be relaxed in the maximization problem (3.17).

Let $\bar{\mathbf{R}}_\theta^* = \{\bar{R}_1, \ldots, \bar{R}_K^*\}$ be the vector of average user spectral efficiencies associated with a given vector of relative priorities $\theta = \{\theta_1, \ldots, \theta_K\}$. Ideally, the priority vector θ could be designed to achieve a desired working point of spectral efficiencies $\bar{\mathbf{R}}_\theta^*$. Unfortunately, no general relationship between θ and $\bar{\mathbf{R}}_\theta^*$ is formulated in the literature and only some particular cases have been presented. The only general interpretation that can be associated with the relative priorities θ_k is that the hyperplane described by $\sum_{k=1}^{K} \theta_k \bar{R}_k = $ constant is tangent to the boundary of the region at point $\{\bar{R}_1^*, \ldots, \bar{R}_K^*\}$ where the weighted sum of average spectral efficiencies is maximized, i.e. $\bar{R}_\theta^* = \max\left(\sum_{k=1}^{K} \theta_k \bar{R}_k\right)$. Typically, there is only one point that maximizes the weighted sum rate. However, in some situations there might be more than one point that maximizes such weighted sum for a given priority vector. In that case the hyperplane is tangent to the boundary of the region in all these points. This is illustrated in Figure 3.2 for the two users case. The hyperplane $\theta_1 \bar{R}_1 + \theta_2 \bar{R}_2 = $ constant is a straight line and depending on the value of the constant, some points of the straight line might be inside the average spectral efficiency region. As the weighted sum of average spectral efficiencies is to be maximized, the constant defining the straight line must be also maximized while at the same time, some points of the straight line must be in the region. Thus, ensuring a feasible solution. By fixing θ_1 and θ_2, the slope of the straight line is also fixed. Then, the constant increases by increasing either $\bar{R}_1$ and keeping $\bar{R}_2$ fixed or vice versa. By increasing the constant, the straight line is moved to the right in Figure 3.2 (left) and the maximum weighted sum is achieved when the constant cannot be further increased by increasing either $\bar{R}_1$ or $\bar{R}_2$ without leaving the region. When this is the case, the straight line is tangent to the boundary of the region. Typically, the tangency

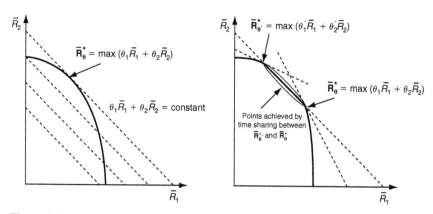

Figure 3.2 *Two-user average spectral efficiency region and maximum weighted sum rate.*

is at only one point, Figure 3.2 (left), but in some cases there is more than one tangent point, Figure 3.2 (right). However, any of these points can be achieved through a time-sharing operation between two other points obtained by changing the priority vector $\boldsymbol{\theta}$.

Figure 3.3 illustrates an arbitrary two-user average spectral efficiency region. The boundary of the region is the curve ABCDEF. The extreme points F and A on this boundary correspond to the single-user average spectral efficiencies denoted by $\bar{R}_1^1$ and $\bar{R}_2^1$ of users 1 and 2, respectively. More precisely, $\bar{R}_1^1$ is $\bar{R}_1^*$ when $\theta_1 = 1$ and $\theta_2 = 0$ and $\bar{R}_2^1$ is $\bar{R}_2^*$ when $\theta_1 = 0$ and $\theta_2 = 1$. In that case, it is easy to observe that the resource allocation policy is the trivial solution to (3.17) given by $\mathbf{p}_{\boldsymbol{\theta}}^*(\mathbf{H}) = \{\theta_1 P, \theta_2 P\}$. Clearly, this is equivalent to the single user scenario presented in Chapter 2. Note that in the example it is assumed that $\bar{R}_1^1 > \bar{R}_2^1$; a reason for that could be that the average SNR (as described in Chapter 2) relative to user 1 is higher than that of user 2.

Point E, with a local tangent at 45 degrees, gives the maximum sum of average spectral efficiencies $\max_{\boldsymbol{\theta}} \left(\bar{R}_1^* + \bar{R}_2^* \right)$. Clearly, such a working point is achieved by setting equal priorities $\theta_1 = \theta_2 = 1$ and applying (3.17). Although this setting maximizes the system performance, it generally results in unfair situations where

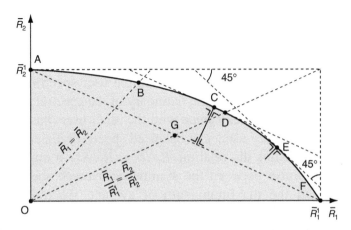

Figure 3.3 *Two-user average spectral efficiency region and specific points of the boundary.*

the users presenting the best average SNR achieve much higher spectral efficiencies than the others, which might not be desirable in many practical applications.

p0400   Point B, on the other hand, gives the maximum common average spectral efficiency or symmetric spectral efficiency. When the single user average spectral efficiencies are very different, setting a common average spectral efficiency constraint for all users is generally a waste of resources as it forces the users with the best average SNR to lower their spectral efficiency dramatically to reach the level of the weakest users.

p0410   The line AGF represents the rate distributions obtained by using time-division multiple-access (TDMA). Spectral efficiencies $1/2\bar{R}_1^1$ and $1/2\bar{R}_2^1$ are obtained if time slots of equal duration are allocated to each user (point G). That is, if the resource allocation policy (3.17) is applied with $\theta_1 = 1$ and $\theta_2 = 0$ half of the time and with $\theta_1 = 0$ and $\theta_2 = 1$ the other half. Note that the time sharing between these two priority vectors can be done either randomly (with equal probability) or deterministically. Because of the priorities associated with TDMA strategy (either $\theta_1 = 1$ and $\theta_2 = 0$ or $\theta_1 = 0$ and $\theta_2 = 1$), it is clear that such strategy is limited to only one transmission at a time and does not use the channel state information **H**.

p0420   Alternatively, one could think of the *balanced spectral efficiency* of a multi-user channel as an example of a specific spectral efficiency distribution that satisfies a fairness criterion. It is defined as the distribution of maximum simultaneously achievable spectral efficiencies that are proportional to the single user spectral efficiencies. It is a specific point on the boundary of the spectral efficiency region for which the coexistence with the other users has the same relative cost for every user. The balanced spectral efficiency, given by point D, satisfies the relation $\bar{R}_1/\bar{R}_1^1 = \bar{R}_2/\bar{R}_2^1$. It appears as a smart compromise between the symmetric average spectral efficiency B and the maximum sum of average spectral efficiencies E. A higher balanced spectral efficiency is achieved at point D with respect to that at point G; this is due to the multiple simultaneous transmissions that are multiplexed with an appropriate power allocation according to **H**. In any case, the maximum balanced rates can be written as

$$\bar{R}_k = g \frac{\bar{R}_k^1}{K} \qquad (3.18)$$

where $g \geq 1$ is the spectral efficiency gain with respect to the TDMA strategy (OD/OG in the figure). Note that the maximum spectral efficiency gain depends on the multiplexing capabilities of the receiver (uplink) or transmitter (downlink). More details on the spectral efficiency gain due to user multiplexing are given in Chapters 5 and 6.

Apart from the working points considered above, a panoply of working points with the consequent resource allocation strategies could be also considered. As, for instance, in a system where data are stored in buffers before being transmitted, it is shown that by setting the user priorities proportional to their buffer sizes the average delay is minimized and the system is kept stable, i.e. buffers do not blow up [5, 6 and 7]. The relationship between resource allocation and delay is studied in more detail in Chapter 7.

Whatever the desired working point, the optimal resource allocation given by (3.17) is to be performed. Therefore, the convexity of the spectral efficiencies $R_k^*(\mathbf{H}, \mathbf{p}(\mathbf{H}))$ with respect to $\mathbf{p}(\mathbf{H})$, for a given ordering vector $\boldsymbol{\pi}$ if a sequential encoder/decoder is used, is of interest so that convex optimization methods can be used to obtain the optimal solution. This will be the main topic in the following chapters where different wireless systems will be considered. Next, the last part of this chapter considers the general case where the spectral efficiency measure is the capacity (i.e. the information theory spectral efficiency when optimal transceivers are used).

3.6 A particular case: the capacity region

At this point it should be clear that the largest spectral efficiency region can be obtained by allowing simultaneous access to the channel of different users. The fundamental trade-off between users' spectral efficiencies is parameterized by two components: (i) the power allocation $\mathbf{p(H)} = \{p_1, \ldots, p_k\}$, which should be performed jointly for all users, and (ii) the encoding/decoding order of users, with $K!$ possible orderings of the users.

p0460

If we particularize the spectral efficiency to the case of capacity, it is well known that dirty paper coding achieves the entire capacity region of the MISO broadcast channel [8]. Finding optimum resource allocations for the MISO broadcast channel turns out to be challenging due to the complicated mathematical structure of the broadcast channel and its non-degradedness caused by the spatial degrees of freedom. However, it was shown that the capacity regions of MISO broadcast channel and its dual SIMO multiple access channel having hermitian transposed channels coincide (as soon as the power constraint is set on the total transmitted power). Further, duality transformations relating the particular capacity achieving power allocations in the broadcast channel to the ones in the multiple access channel were provided so that all problems can be solved in the dual SIMO multiple access channel which has a more favorable structure [2]. If we concentrate on the multiple access channel, we know that an SIC receiver achieves capacity. Therefore, given a sequential decoding order $\pi = \{\pi_1, \ldots, \pi_K\}$ such that user k with $\pi_i = k$ is the ith user to be decoded, the kth user capacity is formulated as

$$
C_k^\pi(\mathbf{H}, \mathbf{p(H)}) = \log_2 \det \left(\mathbf{I} + \frac{1}{\sigma^2} \sum_{\substack{k' \in \mathbb{K} \\ k' \in \{\pi_{i+1}, \ldots, \pi_K\}}} p_{k'} \mathbf{h}_{k'} \mathbf{h}_{k'}^H \right)
$$

$$
- \log_2 \det \left(\mathbf{I} + \frac{1}{\sigma^2} \sum_{\substack{k' \in \mathbb{K} \\ k' \in \{\pi_1, \ldots, \pi_{i-1}\}}} p_{k'} \mathbf{h}_{k'} \mathbf{h}_{k'}^H \right) \quad (3.19)
$$

where $\det(\cdot)$ is the determinant and $\mathbf{I}$ is the identity matrix. The global capacity region is then generated by the union of all the possible orderings and the union over all possible power allocation techniques

$$C(\mathbf{H}, \mathbf{p}(\mathbf{H})) = \bigcup_{\pi \in \Pi} \bigcup_{\mathbf{P} \in \mathbb{P}} \left\{ \begin{array}{l} \mathbf{R}(\mathbf{H}, \mathbf{p}(\mathbf{H})) \mid R_k(\mathbf{H}, \mathbf{p}(\mathbf{H})) \\ \qquad \leq C_k^{\pi}(\mathbf{H}, \mathbf{p}(\mathbf{H})) \end{array} \right. \quad (3.20)$$

When power constraints are individual power constraints (multiple access channel), such capacity region is a polyhedron with $K!$ vertices in the positive quadrant. Each vertex is achievable by a successive encoding/decoding using one of the $K!$ possible orderings of the users. In the special case $K = 2$, the spectral efficiency region is the well-known Cover–Wyner pentagon, with two useful vertices [11]. When power constraints are on the total power consumption, the resulting boundary of the global capacity region is curved.

The ergodic (or average) capacity region of (3.20) is defined by (3.15) and (3.16) with $R_k(\mathbf{H}, \mathbf{p}(\mathbf{H})) = C_k^{\pi}(\mathbf{H}, \mathbf{p}(\mathbf{H}))$, given in (3.19). Each point at the boundary is achieved by the weighted sum maximization in (3.17). For the multiple access channel, (3.17) is a convex optimization problem with optimal ordering solution π given by [10]

$$\theta_{\pi_1} \leq \theta_{\pi_2} \leq \cdots \leq \theta_{\pi_K} \quad (3.21)$$

This optimal ordering solution says that the user with higher priority is the last one to be decoded. Note that the ordering solution depends neither on the channel realization nor on the optimal power allocation $\mathbf{p}_{\theta}^*(\mathbf{H})$. With the optimal ordering solution, the weighted sum in (3.17) can be easily expressed as

$$\sum_{k=1}^{K} \theta_k R_k(\mathbf{H}, \mathbf{p}(\mathbf{H}))$$

$$= \sum_{i=1}^{K} \theta_{\pi_i} \log_2 \det\left(\mathbf{I} + \frac{1}{\sigma^2} \sum_{l=i}^{K} P_{\pi_l} \mathbf{h}_{\pi_l} \mathbf{h}_{\pi_l}^{H} \right)$$

$$\quad - \theta_{\pi_i} \log_2 \det\left(\mathbf{I} + \frac{1}{\sigma^2} \sum_{l=i+1}^{K} P_{\pi_l} \mathbf{h}_{\pi_l} \mathbf{h}_{\pi_l}^{H} \right)$$

$$= \sum_{i=1}^{K} \left(\theta_{\pi_i} - \theta_{\pi_{i-1}} \right)$$

$$\log_2 \det\left(\mathbf{I} + \frac{1}{\sigma^2} \sum_{l=i}^{K} P_{\pi_l} \mathbf{h}_{\pi_l} \mathbf{h}_{\pi_l}^{H} \right) \quad (3.22)$$

with $\theta_{\pi_0} = 0$ by definition. Note that this ordering is crucial because as a result, (3.22) is concave with respect to $\mathbf{p(H)}$ and its maximization can be solved by classical convex optimization techniques.

p0500   Unfortunately, the weighted sum maximization (3.17) is not a convex optimization problem for a large set of transmitters/receivers and therefore a general cross-layer solution does not exist but optimal solutions become particular to each study case. For instance, in the particular case of using a ZF receiver, the optimization problem becomes convex because interferences are nulled at the expense of increasing the noise power. The sequel to this book is devoted to analyzing the average spectral efficiency region together with optimal and suboptimal resource allocation policies and the convexity of (3.17) for different study cases such as SISO, SIMO and MISO scenarios.

s0080   ## 3.7 Summary

p0510   The cross-layer resource allocation problem in multi-user scenarios has been explained in this chapter. We have observed that in a multi-user scenario, the spectral efficiency region appears as a natural extension of the concept of spectral efficiency of the point-to-point scenario presented in the previous chapter. In such a multi-user context, to compute the spectral efficiency, the SNIR accounting for the users' interferences is to be considered instead of the SNR of the point-to-point link. Therefore, a general formulation of the SNIR has been presented.

p0520   The average spectral efficiency region is defined as the set of all the individual average spectral efficiencies that can be achieved by any resource allocation policy and it is shown that any working point in that region can be achieved by allocating resources according to the resource allocation maximization problem (3.17). In the particular case when the spectral efficiency is the channel capacity, some properties of the capacity region are explained and the resource allocation maximization problem becomes a convex optimization problem that can be solved through standard convex optimization methods. However, when this is not the case, the convexity of (3.17) is to be studied and particular solutions are to be found.

3.8 Annex: Examples of signal processing techniques

The capability of the system to deal with interferences mainly depends on the signal processing matrix $\mathbf{V}$ applied either at the transmitter side or at the receiver side. As already mentioned, this signal processing can be applied either at the receiver, so multiple packets are simultaneously received in the uplink, or at the transmitter, so multiple packets are simultaneously transmitted to multiple receivers in the downlink.

3.8.1 The bank of matched filters

Let us consider the simplest signal processing technique such that the signal at the receiver is processed by a bank of match filters, that is, $\mathbf{V} = \mathbf{H}^H$. In that case, the signal model for the multi-user SIMO multiple access channel is

$$
y_k^{MF} = \mathbf{h}_k^H \mathbf{h}_k s_k + \sum_{\substack{k' \in \mathbb{K} \\ k' \neq k}} \mathbf{h}_k^H \mathbf{h}_{k'} s_{k'} + \mathbf{h}_k^H w \tag{3.23}
$$

and the SNIR is easily computed as

$$
\gamma_k^{MF} = \frac{\left| \mathbf{h}_k^H \mathbf{h}_k \right|^2 p_k}{\left| \mathbf{h}_k^H \mathbf{h}_k \right|^2 \sigma^2 + \sum_{\substack{k' \in \mathbb{K} \\ k' \neq k}} \left| \mathbf{h}_k^H \mathbf{h}_{k'} \right|^2 p_{k'}}
$$

$$
= \frac{p_k}{\sigma^2 + \sum_{\substack{k' \in \mathbb{K} \\ k' \neq k}} \frac{\left| \mathbf{h}_k^H \mathbf{h}_{k'} \right|^2}{\left| \mathbf{h}_k^H \mathbf{h}_k \right|^2} p_{k'}} \tag{3.24}
$$

Comparing expressions (3.6) and (3.8) with (3.23), it can be easily shown that (3.24) is also valid in the multi-user MISO broadcast channel because $\left| \mathbf{h}_k^H \mathbf{h}_{k'} \right|^2 = \left| \mathbf{h}_{k'}^H \mathbf{h}_k \right|^2$.

3.8.2 The zero forcing (ZF) beamformer

Instead of a bank of matched filters, a more engineering solution to deal with interferences is to make an interference cancellation among the users through a ZF equalizer. Assume then that during transmission of packets, a ZF beamformer is performed by means of the M antennas. That is, $\mathbf{V} = (\mathbf{H}_{\mathbb{K}}^H \mathbf{H}_{\mathbb{K}})^{-1} \mathbf{H}_{\mathbb{K}}^H$. Note that due to the constraints on the ZF beamformer, successful reception of packets is only possible when $\mathbb{K} \leq M$, so the number of users cannot be higher than the number of antennas. In this case, the received symbol vector is given in matrix form by

$$\mathbf{y}_{\mathbb{K}}^{ZF} = (\mathbf{H}_{\mathbb{K}}^H \mathbf{H}_{\mathbb{K}})^{-1} \mathbf{H}_{\mathbb{K}}^H \mathbf{r}_{\mathbb{K}} \tag{3.25}$$

where $\mathbf{y}_{\mathbb{K}}^{ZF}$ and $\mathbf{r}_{\mathbb{K}}$ are the $\mathbf{y}$ and $\mathbf{r}$ vectors with elements corresponding to users in the set $\mathbb{K}$ and $\mathbf{H}_{\mathbb{K}}$ is the channel matrix containing the columns of $\mathbf{H}$ corresponding to users in the set $\mathbb{K}$. In this scenario no interference is present but the SNIR expression moves back to an SNR expression with a noise enhancement factor. Operating the SNR can be easily shown to be

$$\gamma_k^{ZF} = \frac{p_k}{\sigma^2} \frac{1}{\left[(\mathbf{H}_{\mathbb{K}}^H \mathbf{H}_{\mathbb{K}})^{-1}\right]_{kk}} \tag{3.26}$$

with $\left[(\mathbf{H}_{\mathbb{K}}^H \mathbf{H}_{\mathbb{K}})^{-1}\right]_{kk}$ defining the kth element of the diagonal of $(\mathbf{H}_{\mathbb{K}}^H \mathbf{H}_{\mathbb{K}})^{-1}$. The factor $\left[(\mathbf{H}_{\mathbb{K}}^H \mathbf{H}_{\mathbb{K}})^{-1}\right]_{kk}$ is typically known as the noise enhancement factor. Recall that when $|\mathbb{K}| > M$, $\gamma_k^{ZF} = 0$.

p0570 Clearly, the ZF beamformer can be applied either at the receiver in the multiple access channel or at the transmitter in the broadcast channel showing identical results.

s0120 ### 3.8.3 The minimum mean square error (MMSE) beamformer

p0580 Although the ZF beamformer eliminates interferences, the noise enhancement factor can represent serious damage in some situations. The solution to the noise enhancement problem comes in the form of a regularized ZF process through the MMSE equalizer.

p0590 Consider the case in which the AP uses the linear MMSE beamformer to process the received signal. Such a beamformer minimizes the normalized MSE. In that case the processing matrix has the form $\mathbf{V} = \mathbf{P}_{\mathbb{K}}^{1/2} \mathbf{H}_{\mathbb{K}}^H (\mathbf{H}_{\mathbb{K}}^H \mathbf{P}_{\mathbb{K}} \mathbf{H}_{\mathbb{K}} + \sigma^2 \mathbf{I})^{-1}$ where the matrix $\mathbf{P}_{\mathbb{K}}$ is a diagonal matrix containing the powers p_k of users in $\mathbb{K}$. Note the existence of the $\sigma^2\mathbf{I}$ in the processing therefore, the use of the term of regularized ZF. The received symbol vector is given in matrix form by

$$\mathbf{y}_{\mathbb{K}}^{MMSE} = \mathbf{P}^{1/2}\mathbf{H}_{\mathbb{K}}^{H}(\mathbf{H}_{\mathbb{K}}^{H}\mathbf{PH}_{\mathbb{K}} + \sigma^2\mathbf{I})^{-1}\,\mathbf{r}_{\mathbb{K}} \qquad (3.27)$$

Operating the previous expression, the SNIR can be shown to be [4]

$$\gamma_k^{MMSE} = p_k\mathbf{h}_k^{H}\left(\sum_{\substack{k'\in\mathbb{K}\\k'\neq k}} p_{k'}\mathbf{h}_{k'}\mathbf{h}_{k'}^{H} + \sigma^2\mathbf{I}\right)^{-1}\mathbf{h}_k \qquad (3.28)$$

As for the ZF beamformer, the MMSE beamformer can be applied either at the receiver side in the multiple access channel or at the transmitter side in the broadcast channel.

References

[1] D. Tse and P. Viswanath, *Fundamentals of Wireless Communication*, Cambridge University Press, 2005.

[2] N. Jindal, S. Vishwanath and A. Goldsmith, On the duality of Gaussian multiple-access and broadcast channels, *IEEE Trans. Information Theory*, Vol. 50, Issue 5, pp. 768–783, May 2004.

[3] M. Costa, Writing on dirty paper, *IEEE Trans. Information Theory*, No. 3, pp. 439–441, May 1983.

[4] D.N.C. Tse and S.V. Hanly, Linear multiuser receivers: effective interference, effective bandwidth and user capacity, *IEEE Trans. Information Theory*, Vol. 45, Issue 2, pp. 641–657, March 1999.

[5] M.J. Neely, E. Modiano and C.E. Rohrs, Power allocation and routing in multi-beam satellites with time varying channels, *IEEE Trans. Networking*, Vol. 11, Issue 1, pp. 138–152, Feb. 2003.

[6] E.M. Yeh and A.S. Cohen, Throughput and delay optimal resource allocation in multiaccess fading channels, *Proc. IEEE International Symposium on Information Theory*, p. 245, June 2003.

[7]   M.J. Neely, Dynamic power allocation and routing for sat-
      ellite and wireless networks with time varying channels,
      Ph.D. dissertation, LIDS, Mass. Inst. Technol., Cambridge,
      MA, 2003.

[8]   H. Weingarten, Y. Steinberg and S. Shamai, The capacity
      region of the Gaussian multiple-input multiple-output broad-
      cast channel, *IEEE Trans. Information Theory*, Vol. 52,
      pp. 3936–3964, 2006.

[9]   M.K. Varanasi, and T. Guess, Optimum decision feedback
      multiuser equalization with successive decoding achieves
      the total capacity of the Gaussian multiple-access channel,
      Proc. Thirty-first Asilomar Conference, Nov. 1997.

[10] H. Boche, A. Jorswieck and T. Haustein, Channel aware
      scheduling for multiple antenna multiple access channels,
      Proc. Thirty-seventh Asilomar Conference, Nov. 2003.

[11] D. Tse and S. Hanly, Multi-access fading channels: Part I:
      Polymatroid structure, optimal resource allocation and
      throughput capacities, *IEEE Trans. Information Theory*,
      Vol. 44, Issue 7, pp. 2796–2815, Nov. 1998.

4

Cross-layer resource allocation in SISO systems

0004

In a wireless network with many users, the variability of the wireless channel creates, at each time instant, different channel conditions for each of the users. This phenomena is known as multi-user diversity (MUD). Therefore, if CSI is available at the resource allocation unit, a MUD gain can be exploited.

Following a PHY layer perspective, MUD has been addressed in many different directions [1–5] investigating ergodic capacity and fairness of MUD resource allocation.

Following a Link/MAC layer perspective, it is considered that data are transmitted in the form of packets and therefore the spectral efficiency is evaluated in terms of average throughput rather than ergodic capacity. Although techniques such as link adaptation might help to reduce the differences between capacity and throughput, in general, information-theoretic spectral efficiencies are not achievable by the PHY layer and there is always a gap between the theoretical achievable spectral efficiency (capacity) and the real achieved one (throughput). This suggests that information-theory approaches to MUD should be revisited under a cross-layer perspective.

In this chapter the MUD problem is approached from a cross-layer perspective considering packet-oriented transmissions and that packet errors occur. Starting from a general power allocation problem, we see that a MUD scheduling policy is a simplified resource allocation policy that allocates time resources according to the users' channel states. We evaluate under what conditions MUD scheduling is optimal and then the average spectral efficiency region of MUD scheduling is studied and differences between the average throughput region and the average capacity region are shown.

p0050   Besides, in this chapter we explain that not only MUD but also other degrees of diversity might appear in multi-user environments. Indeed, due to the great advances and rapid development of wireless communications, wireless networks are evolving to networks with heterogeneity. Networks with heterogeneity among users include scenarios where: (i) users transmit packets corresponding to different services (voice, data, etc.) with different packet lengths, (ii) single antenna users share channel resources with multiple antenna users, (iii) users with legacy terminals (e.g. IEEE802.11b) live together with users with advanced terminals (e.g. IEEE802.11g) or (iv) users are multi-standard users. Consequently, each user has its particular way of exploiting channel conditions leading to an heterogeneous multi-user diversity (HMUD). In this chapter, the average throughput is also investigated in networks with HMUD.

s0010   ## 4.1   MUD scheduling: the optimal policy

p0060   Consider either the downlink or the uplink of a centralized wireless network with K single antenna mobile users that have to be served by a single antenna access point (AP). See Figure 4.1 for an example of such a network. A perfect instantaneous feedback channel is considered for enhancing the downlink. By means of the messages on the feedback channel, the AP acquires knowledge of the user's channel state required for implementing MUD scheduling. Conversely, in the uplink, the AP gives access to the mobile users by means of a polling mechanism. Then, a mobile user can only transmit upon a previous reception of a polling packet. A perfect instantaneous polling channel is assumed. Notice that whether the communication is in the uplink or in the downlink, once the AP has knowledge of the channel estimates, MUD scheduling consists of mapping such channel estimates into a quality level (according to a defined spectral efficiency function) and on allocating resources according to the quality level computed.

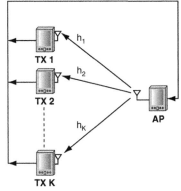

Figure 4.1   *Multi-user SISO wireless network.*   f0010

070 Our resource allocation will be in the form of (3.17). Let us consider that we want to maximize the sum of the average spectral efficiencies. Then, from Chapter 3 we know that this is obtained when all users have the same priority. Further, if the resource allocation policy is approached from an information-theory perspective, the spectral efficiency is the capacity. In consequence, expression (3.22) in Chapter 3 can be used and, for all users having the same priority, the sum capacity is reduced to

$$\sum_{k=1}^{K} \theta_k R_k(\mathbf{h}, \mathbf{p}(\mathbf{h})) = \sum_{k=1}^{K} R_k(\mathbf{h}, \mathbf{p}(\mathbf{h}))$$
$$= \log_2 \left(1 + \frac{1}{\sigma^2} \sum_{k=1}^{K} p_k |h_k|^2 \right) \qquad (4.1)$$

080 Note that in the SISO case, the channel matrix $\mathbf{H}$ is reduced to a $K \times 1$ channel vector $\mathbf{h}$ where the kth element is the kth user SISO channel. Following (3.17), the optimal resource allocation policy $\mathbf{p}_\theta^*(\mathbf{h})$ that maximizes the sum of ergodic individual capacities is such that

$$\mathbf{p}_\theta^*(\mathbf{h}) = \arg \max_{\mathbf{p}(\mathbf{h})} \left(\log_2 \left(1 + \frac{1}{\sigma^2} \sum_{k=1}^{K} p_k |h_k|^2 \right) \right) \qquad (4.2)$$

090 subject to given power constraints.

4.1.1 *Average power constraints*

Let us consider the downlink communication scenario with average power constraints $E\left\{\sum_k p_k\right\} \le P_{total}$. Then, the maximization problem (4.2) is solved by means of classical convex optimization methods [6] as follows.

The Lagrange dual function of the maximization problem (4.2) is given by

$$\mathcal{L}(\mathbf{p}(\mathbf{h}), \lambda) = -\log_2\left(1 + \frac{1}{\sigma^2}\sum_{k=1}^{K} p_k |h_k|^2\right)$$

$$+ \lambda\left[E\left\{\sum_k p_k\right\} - P_{total}\right] - \sum_k \mu_k p_k \qquad (4.3)$$

where λ is the Lagrange multiplier for the total power restrictions and μ_k are the Lagrange multipliers for the intrinsic individual power restrictions $p_k \ge 0$. The first derivative of the Lagrangian with respect to p_k is

$$\frac{\partial\mathcal{L}(\mathbf{p}(\mathbf{h}), \lambda)}{\partial p_k} = -\frac{|h_k|^2}{\sigma^2 + \sum_{k=1}^{K} p_k |h_k|^2} + \lambda - \mu_k \qquad (4.4)$$

The Karush–Kuhn–Tucker (KKT) conditions for this optimization problem are necessary and sufficient for the optimality of the power allocation $\mathbf{p}(\mathbf{h})$. From (4.4), we have the following KKT conditions

$$\frac{|h_k|^2}{\sigma^2 + \sum_{k=1}^{K} p_k |h_k|^2} = \lambda - \mu_k \qquad 1 \le k \le K$$

$$p_k \ge 0 \qquad 1 \le k \le K$$

$$\mu_k \ge 0 \qquad 1 \le k \le K$$

$$\mu_k p_k = 0 \qquad 1 \le k \le K$$

$$E\left\{\sum_k p_k\right\} - P_{total} \le 0$$

$$\lambda \geq 0$$

$$\lambda \left(E \left\{ \sum_k p_k \right\} - P_{total} \right) = 0 \qquad (4.5)$$

Assuming that h_k are all different, from the first condition we observe that only one user can be active at a time. That is, power is allocated to only one user. Furthermore, we observe that (4.1) is maximized when the selected user is the one with maximum $|h_k|^2$. Then, the optimal power allocation $\mathbf{p}_\theta^*(\mathbf{H}) = \{ p_1^*, ..., p_K^* \}$ when $\theta_k = 1$ for $1 \leq k \leq K$, is given by

$$p_k^* = \begin{cases} \left(\dfrac{1}{\lambda} - \dfrac{\sigma^2}{|h_k|^2} \right)^+ & \text{if } k = \arg \max_{k'} |h_{k'}|^2 \\ 0 & \text{otherwise} \end{cases} \qquad (4.6)$$

where $(x)^+$ denotes $\max(0, x)$ and the parameter λ must be designed such that the average power constraints are satisfied. Note that p_k^* maximizes the sum capacity by choosing the user with the best channel state and allocating power to that user. In particular, from (4.6) we observe that more power is allocated in good channels than in bad channels.

In the uplink, with individual average power constraints $E\{p_k\} \leq P$ for $1 \leq k \leq K$, similar conclusions are obtained [7]. To illustrate this fact, note that with the resource allocation policy (4.6), by symmetry, the power consumption of all users is the same. Consequently, individual power constraints with $p_k = P_{total}/K$ are automatically satisfied.

In conclusion, either in the uplink or in the downlink but with average power constraints, the optimal resource allocation that maximizes the sum capacity is to allocate resources to the best user.

4.1.2 Instantaneous power constraints

The case for instantaneous power constraints is different for the downlink and for the uplink. In the downlink, the first derivative of the Lagrangian of the maximization problem (4.2) is the same as (4.4). Therefore, the optimal solution is the same as (4.6) but with parameter λ designed such that the instantaneous power constraints are satisfied. In other words, the optimal solution is

$$
p_k^* = \begin{cases} P_{total} & \text{if } k = \arg\max_{k'} |h_{k'}|^2 \\ 0 & \text{otherwise} \end{cases} \tag{4.7}
$$

p0170 The optimal power allocation (4.7) says that the optimal resource allocation in the downlink is to allocate all power to the user with the best channel state.

p0180 However, in the uplink, due to the individual power constraints, the Lagrangian is

$$
\mathcal{L}(\mathbf{p}(\mathbf{H}), \lambda) = -\log_2\left(1 + \frac{1}{\sigma^2}\sum_{k=1}^{K} p_k |h_k|^2\right)
$$
$$
+ \lambda_k(p_k - P) - \sum_k \mu_k p_k \tag{4.8}
$$

p0190 Leading to the following KKT conditions

$$
\frac{|h_k|^2}{\sigma^2 + \sum_{k=1}^{K} p_k |h_k|^2} = \lambda_k - \mu_k \qquad 1 \leq k \leq K
$$
$$
\begin{aligned}
p_k &\geq 0 & 1 \leq k \leq K \\
\mu_k &\geq 0 & 1 \leq k \leq K \\
\mu_k p_k &= 0 & 1 \leq k \leq K \\
p_k - P &\leq 0 & 1 \leq k \leq K \\
\lambda_k &\geq 0 & 1 \leq k \leq K \\
\lambda_k(p_k - P) &= 0 & 1 \leq k \leq K
\end{aligned} \tag{4.9}
$$

and the optimal solution $p_k^* = P$ is obtained immediately. This solution implies that all users transmit simultaneously at maximum power.

p0200 Conclusions as to whether only one user or all users can transmit could be obtained directly from taking a careful look at expression

(4.1). Clearly, when considering restrictions on the total power consumption (downlink), expression (4.1) is maximized by scheduling the user k such that $|h_k|^2 = \max_i |h_i|^2$. The grounds for this is that there is no reason why power cannot be allocated to another user if there is a user that can better exploit such amount of power. The amount of power to be allocated to this user will then depend on whether restrictions are on average, given by (4.6), or are instantaneous, given by (4.7).

210 On the other hand, when working with individual power constraints (uplink), one could consider allocating power to more than one user. We observe that the increase in the sum capacity (4.1) for a given amount of power allocated to one user is higher when the argument of the logarithm is small. Therefore, when restrictions are on average, it is more efficient to allocate power to only one user and reserve the power of other users for another time instant. The reason why the user with the best channel is chosen is because users with bad channels can wait until they experience a good channel. Because restrictions are on average, this is permissible. The amount of power to be allocated to a user is given by (4.6). However, in the case of instantaneous individual power restrictions, the power that is not allocated to one user at a given time instant is lost. Therefore, the most efficient solution in this case is to allow all users to transmit at maximum power.

220 The simplest power allocation solution is (4.7), that is, to allocate all the power to the best user. In fact, this has been the most extensive solution considered in practice because the power allocation problem is converted into a simpler scheduling problem where the best user is to be scheduled. This strategy is known as MUD scheduling or opportunistic scheduling. Even when working with individual instantaneous power constraints in the uplink, where we showed that this is not the optimal solution, the MUD scheduling strategy is commonly used in practice because of its simplicity.

230 Since sum capacity gains of MUD scheduling have been extensively studied, rather than presenting another information-theoretic approach to MUD scheduling, the motivation for the rest of this chapter is to present a cross-layer study of MUD scheduling.

s0040   ## 4.2 Cross-layer approach to MUD scheduling

p0240   In the previous sections, we showed that the MUD scheduling policy, i.e. the policy that allocates resources to the best user, maximizes the sum capacity in the downlink and the sum capacity in the uplink when average power constraints are imposed. Interestingly enough is the fact that the sum capacity is achieved without the need for iterative processing techniques (SIC or DPC). However, a more complex power allocation solution is obtained when, instead of the sum capacity, the weighted sum capacity is to be maximized [8, 9]. In this case, transceivers with iterative processing capabilities are necessary to achieve capacity. This is also the case for the sum capacity maximization in the uplink with individual instantaneous power constraints.

p0250   When ideal transceivers are not used but instead more practical transceivers are considered, the channel capacity is no longer achievable. Then, the concept of spectral efficiency as explained in Chapters 2 and 3 comes into play. In these chapters we defined the information-theory rates as the ideally achievable rates when the SNIR is determined by a given transceiver architecture and throughput was defined as the error-free rate at the MAC/Link layer, not only when the SNIR is determined by a given transceiver architecture but also when particular modulation and coding schemes are considered. In these cases, it can happen (as already highlighted in Chapter 3, sections 3.5 and 3.6) that the maximization problem (3.17) is no longer a convex optimization problem.

p0260   In summary, MUD scheduling might not be the optimal solution when the weighted sum of spectral efficiencies is to be maximized. Nevertheless, with the aim of presenting a cross-layer analysis of resource allocation rather than presenting optimal but complex resource allocation techniques, we will study MUD scheduling from a cross-layer perspective even though MUD scheduling might not always be the optimal solution.

s0050   ### 4.2.1 The average spectral efficiency region of MUD scheduling

p0270   Following the framework presented in Chapter 3, we define the average spectral efficiency region as

$$\Omega = \left\{ \sum_{k=1}^{K} \theta_k \bar{R}_k \mid \theta_k \leq 1, \sum_{k=1}^{K} \theta_k = 1, \bar{R}_k \in S, k \in \{1, ..., K\} \right\} \quad (4.10)$$

where

$$S = \bigcup_{\mathbf{P} \in \mathbb{P}} \left\{ \bar{\mathbf{R}} : \bar{R}_k = E\{R_k(\mathbf{h}, \mathbf{p}(\mathbf{h}))\}, \ k \in \{1, ..., K\} \right\} \quad (4.11)$$

According to the description in Chapter 3, any point at the boundary of Ω is achieved by a vector of relative priorities $\boldsymbol{\theta} = \{\theta_1, ..., \theta_K\}$ and a resource allocation policy described by the vector $\mathbf{p}_{\boldsymbol{\theta}}^*(\mathbf{h})$ given by

$$\mathbf{p}_{\boldsymbol{\theta}}^*(\mathbf{h}) = \arg\max_{\mathbf{p}(\mathbf{h})} \left(\sum_{k=1}^{K} \theta_k R_k(\mathbf{h}, \mathbf{p}(\mathbf{h})) \right) \quad (4.12)$$

Previously, we defined the MUD scheduling policy as a power allocation policy that allocates all power to only one user at a time. Under these restrictions, it is easy to see that the optimal power allocation (4.12) is given by

$$p_k^{MUD} = \begin{cases} P & \text{if } k = \arg\max_{k'} \theta_{k'} R_{k'}(\mathbf{h}, \mathbf{1}_{k'}(P)) \\ 0 & \text{otherwise} \end{cases} \quad (4.13)$$

where the indicator vector $\mathbf{1}_k(P)$ is the power allocation vector $\mathbf{p}(\mathbf{h})$ with all components equal to 0 and the kth component equal to P. Clearly, because all the power is allocated to the best user, the resource allocation problem is no longer a power allocation problem but a scheduling problem.

When maximizing the sum capacity in the previous section, we observed that MUD scheduling was characterized by choosing the user with the best channel state but now, with spectral efficiencies defined as $R_k(\mathbf{h}, \mathbf{p}(\mathbf{h})) = \log_2(1 + \gamma_k)$ or $R_k(\mathbf{h}, \mathbf{p}(\mathbf{h})) = R^{(m^*)} PSR^{(m^*)}(\gamma_k)$ and the priority vector $\boldsymbol{\theta} = \{\theta_1, ..., \theta_K\}$, the cross-layer MUD scheduling policy has been defined as a policy that chooses the user with the best weighted spectral efficiency. If we consider throughput as our figure of merit, the region Ω is the average throughput region and hence, particularizing MUD scheduling policy (4.13) to throughput, we obtain

$$p_k^{MUD} = \begin{cases} P & \text{if } k = \arg\max_{k'} \theta_{k'} R^{(m^*)} PSR^{(m^*)}(\gamma_{k'}) \\ 0 & \text{otherwise} \end{cases} \quad (4.14)$$

where the dependence of the SNR on $\mathbf{p}(\mathbf{h})$ is given by the fact that the power allocation vector is of the form $\mathbf{1}_k(P)$ and hence,

$$\gamma_k = \frac{P|h_k|^2}{\sigma^2}$$

p0290 Obviously, if instead of considering throughput, we consider information-theory rate as our figure of merit, the optimal policy is

$$p_k^{MUD} = \begin{cases} P & \text{if } k = \arg\max_{k'} \theta_{k'} \log_2(1 + \gamma_{k'}) \\ 0 & \text{otherwise} \end{cases} \tag{4.15}$$

p0300 and then an information-theory rate region Ω' is obtained.

s0060 ## 4.2.2 An example of average spectral efficiency region in homogeneous networks

p0310 In Figure 4.2 we show the differences between the regions Ω and Ω' in a two-user network. Clearly, $\Omega \subseteq \Omega'$ because information-theory rates are not achievable at the MAC/Link layer. The boundary of the regions is obtained by simulating the MUD scheduling policy for many different values of $\boldsymbol{\theta} = \{\theta_1, ..., \theta_K\}$. The average throughput region obtained by the TDMA (round robin) scheduling policy has also been plotted for comparison. Hence, the advantages of MUD scheduling with respect to TDMA (round robin) are clearly shown in the figure. These advantages are due to the fact that MUD scheduling schedules users using CSI.

p0320 The sum of individual spectral efficiencies is maximized when $\theta_k = 1$ for $k = 1, ..., K$. Then, the MUD scheduling policy finds the user such that $\arg\max_k \{\log_2(1 + \gamma_k)\}$ or $\arg\max_k \{R^{(m^*)} PSR^{(m^*)}(\gamma_k)\}$. The average spectral efficiencies achieved in this case correspond to the points when $(\theta_1, \theta_2) = (1, 1)$ in Figure 4.2. Note that in this case, the equality $\arg\max_k \{\log_2(1 + \gamma_k)\} = \arg\max_k \{\gamma_k\}$ always holds and the equality $\arg\max_k \{R^{(m^*)} PSR^{(m^*)}(\gamma_k)\} = \arg\max_k \{\gamma_k\}$ holds when all users are homogeneous, i.e. all use the same packet lengths and the same set of modulation and coding schemes. Therefore, when aiming at the maximum sum of spectral efficiencies, the MUD scheduling policy is simplified to choose the user with the highest SNR. However, the inequality $\arg\max_k \{\theta_k R^{(m^*)} PSR^{(m^*)}(\gamma_k)\} \neq \arg\max_k \{\gamma_k\}$ holds in general. Then, the SNR information is not

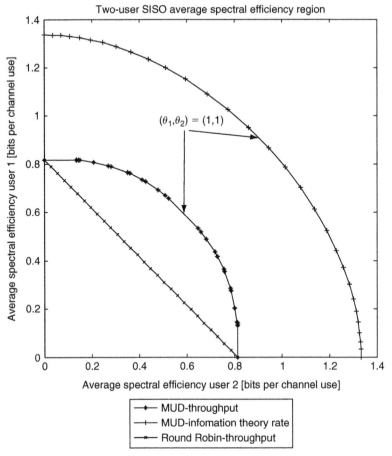

Figure 4.2 *Average spectral efficiency region for MUD scheduling in an SISO system with two users. Rayleigh fading channel with average SNR = 3 dB. Throughput is computed considering that both users use the eight modulation and coding schemes with packet length equal to 1152 information bits as presented in Table 2.3 in section 2.7.*

enough for the correct implementation of MUD scheduling. Indeed, apart from the users' priorities, additional cross-layer information regarding modulation, coding schemes and packet lengths is also necessary.

4.2.3 An example of average spectral efficiency region in heterogeneous networks

In wireless heterogeneous networks users exploit channel conditions in many different ways, and then the user that presents the highest SNR is not necessarily the one that maximizes the sum of individual throughputs. Therefore, in heterogeneous networks, the inequality $\arg \max_k \left\{ \theta_k R^{(m*)} PSR^{(m*)}(\gamma_k) \right\} \neq \arg \max_k \left\{ \gamma_k \right\}$ holds even when $\theta_k = 1$ for $k = 1,\dots, K$.

p0340   The spectral efficiency region of a two-user heterogeneous network is shown in Figure 4.3. In this case, heterogeneity comes from the fact that user 1 transmits using eight modulation and coding schemes and user 2 transmits using only four modulation and coding schemes. We observe that the average information-theory rate region is symmetric, meaning that no differences among

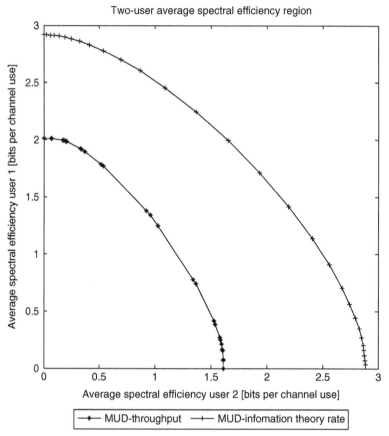

f0030   **Figure 4.3** *Average spectral efficiency region for MUD scheduling in an SISO system with two users using different modulation and coding schemes. Rayleigh fading channel with average SNR = 10 dB. The modulation and coding schemes presented in Table 2.3 in section 2.7 are considered.*

users are experimented when studying information-theory rates. However, the average throughput region is not symmetric. Hence, differences among users are identified by analyzing throughput. Such differences among users are due to what we define as heterogeneous multi-user diversity (HMUD). Note that this example could represent the case of one user with an advanced terminal sharing resources with another user with a legacy terminal. Therefore, a similar case would be a multiple antenna terminal sharing resources with a single antenna terminal making the concept of HMUD even more interesting and to be potentially studied in future systems.

4.3 Total average throughput in networks with HMUD

In the previous section, the effects of HMUD on the average throughput region were shown. The HMUD appears when terminals exploit the wireless channel in different ways. Although heterogeneity among terminals might come from many different causes, the general conclusion is that the knowledge of the users' SNR at the scheduler is not enough to efficiently allocate resources but the cross-layer information on how such SNR values impact on the performance of the users is also necessary.

The average throughput region was studied through the maximization of the weighted sum of individual throughputs. Then, we showed that the system can work at any point on the boundary of the average throughput region by modifying the priority vector θ. However, given a vector of relative priorities, neither the individual average throughput experienced by a user nor the sum of all the individual average throughputs was studied but, from previous chapters, we only know that the sum of individual average throughputs is maximized when all users have the same priority. The sum of all the individual average throughputs is called the total average throughput. In this section, we study the performance of the system in terms of the total average throughput in networks with HMUD.

As a simple example, consider two users denoted by k and k'. User k is transmitting data using a fixed modulation and coding scheme

m and is scheduled with relative priority θ_k, and user k' is transmitting its data using a fixed modulation and coding scheme m' and is scheduled with relative priority $\theta_{k'}$. According to (4.14), the MUD scheduler chooses user k as long as

$$\theta_k R^{(m)} PSR^{(m)}(\gamma_k) > \theta_{k'} R^{(m')} PSR^{(m')}(\gamma_{k'}) \qquad (4.16)$$

Operating the previous expression we obtain that throughput of user k is higher than throughput of user k' as long as the SNR of user k' is below a threshold $\gamma_{k'}^{th}$ given by

$$\gamma_{k'} < \gamma_{k'}^{th} = \left(PSR^{(m')} \left(\frac{\theta_k R^{(m)} PSR^{(m)}(\gamma_k)}{\theta_{k'} R^{(m')}} \right) \right)^{-1} \qquad (4.17)$$

Note that the PSR is always equal to or less than one. Therefore, a value of $\gamma_{k'}^{th}$ always exists if the argument $\theta_k R^{(m)} PSR^{(m)}(\gamma_k)/\theta_{k'} R^{(m')}$ is less than one or equivalently, if $\theta_{k'} R^{(m')} > \theta_k R^{(m)}$. Otherwise, $\gamma_{k'}^{th}$ might tend to infinity for some values of γ_k.

p0380   Now, we proceed to evaluate what is the total average throughput when in the wireless network HMUD is given by the fact that there are K users as user k and K' users as user k'. That is, when there are K users using a fixed modulation and coding scheme m and a relative priority θ_k, and K' users using a fixed modulation and coding scheme m' and a relative priority $\theta_{k'}$. Assume that the users' SNR are i.i.d. random variables with c.d.f. $F(\gamma)$ and p.d.f. $f(\gamma)$.

p0390   Since all the K users use the same modulation and coding scheme and have the same priority, the user with the maximum SNR among the K users is also the user that experiences maximum throughput among these users. The same applies among the K' users. However, a user k belonging to the K users will only be the one with maximum throughput among the K' users if all the K' users experience an SNR below $\gamma_{k'}^{th}$ given by (4.17). Therefore, the MUD scheduler (4.14) schedules user k if all the K' users experience an SNR below $\gamma_{k'}^{th}$ and the SNR of all the K users, except user k, is below the SNR of user k, γ_k.

p0400   In consequence, using basic order statistics, the kth user average throughput is easily formulated as

$$\bar{R}_k = R^{(m)} \int_0^\infty PSR^{(m)}(\gamma_k)(F(\gamma_k))^{K-1}(F(\gamma_{k'}^{th}))^{K'} f(\gamma_k) d\gamma_k \quad (4.18)$$

and the total average throughput is the sum over all individual average throughputs. In (4.18), the throughput of user k is averaged over the SNR but conditioned to the fact that user k is scheduled, i.e. is the best user. Then, $(F(\gamma_{k'}^{th}))^{K'}$ is the probability of user k being the best among the K' users and $(F(\gamma_k))^{K-1}$ is the probability that user k is the best among the K users.

Expression (4.18) is the individual average throughput given user priorities θ_k for $k = 1, \ldots, K$ and $\theta_{k'}$ for $k' = 1, \ldots, K'$. In general, the individual average throughput (4.18) must be evaluated either numerically or by simulations. However, using the well-known step function approximation of the PSR, which is a good approximation in the case of having QoS constraints, and assuming a Rayleigh fading channel with average SNR denoted by $\bar{\gamma}$, i.e. $F(\gamma_k) = 1 - e^{-\gamma_k/\bar{\gamma}}$, a closed form average throughput approximation can be obtained.

First, let us recall that the step function approximation of the PSR was given in Chapter 2 and is expressed as

$$PSR^{(m)}(\gamma) = \begin{cases} 0 & \text{for } \gamma \leq \gamma^{(m)} \\ 1 & \text{for } \gamma \geq \gamma^{(m)} \end{cases} \quad (4.19)$$

Then, condition (4.17) is directly

$$\gamma_{k'} < \gamma_{k'}^{th} = \begin{cases} \gamma^{(m')} & \text{if } \theta_{k'} R^{(m')} > \theta_k R^{(m)} \\ \infty & \text{otherwise} \end{cases} \quad (4.20)$$

and the probability that all the K' users experience an SNR below $\gamma_{k'}^{th}$ is

$$(F(\gamma_{k'}^{th}))^{K'} = \left(1 - e^{-\frac{\gamma_{k'}^{th}}{\bar{\gamma}}}\right)^{K'} \quad (4.21)$$

Besides, the probability that the SNR of user k is the maximum among all K users is

$$(F(\gamma_k))^{K-1} = \left(1 - e^{-\frac{\gamma_k}{\bar{\gamma}}}\right)^{K-1} = \sum_{k=0}^{K-1} \binom{K-1}{k}(-1)^k e^{-k\frac{\gamma_k}{\bar{\gamma}}} \quad (4.22)$$

where we used $(1-x)^q = \sum_{k=0}^{q} \binom{q}{k} x^k (-1)^k$ when $|x| < 1$.
Finally, the average throughput is computed as

$$
\bar{R}_k = R^{(m)} \int_{\gamma^{(m)}}^{\infty} \left(1 - e^{-\frac{\gamma_{k'}^{th}}{\bar{\gamma}}}\right)^{K'} \left[\sum_{k=0}^{K-1} \binom{K-1}{k} (-1)^k e^{-k\frac{\gamma_k}{\bar{\gamma}}}\right] e^{-\frac{\gamma_k}{\bar{\gamma}}} d\gamma_k
$$

$$
= R^{(m)} \left(1 - e^{-\frac{\gamma_{k'}^{th}}{\bar{\gamma}}}\right)^{K'} \sum_{k=0}^{K-1} \binom{K-1}{k} (-1)^k \left(\frac{1}{k+1}\right) e^{-(k+1)\frac{\gamma^{(m)}}{\bar{\gamma}}} \quad (4.23)
$$

Expression (4.23) is a closed form approximation of the individual average throughput obtained when MUD scheduling (4.14) is considered in a scenario with K' users using a modulation and coding scheme denoted by m' and with K users using a modulation and coding scheme denoted by m. Note that the value of $\gamma_{k'}^{th}$ depends, among others, on the vector of relative priorities. The total average throughput is the sum of the individual average throughputs of all the $K + K'$ users.

p0430   In Figure 4.4 an example of the total average throughput computed as the sum of the individual throughputs (4.23) is plotted. It is assumed that the total number of users $K + K'$ is 10 but the number of users K' changes from 0 to 10. Therefore, the total average throughput is computed for different scenarios starting from a scenario with $K = 10$ and $K' = 0$ users and ending with a scenario with $K = 0$ and $K' = 10$.

p0440   In the figure we assumed $\theta_k = \theta_{k'} = 1$ for $k = 1, \ldots, K$ and $k' = 1, \ldots, K'$ but results for other priority vectors could be obtained also. For the sake of comparison, the total average throughput of the MUD policy that chooses users as if they were homogeneous is also plotted. If all users have the same relative priority and users are treated as homogeneous, the MUD policy chooses the user with maximum SNR. Therefore, the individual

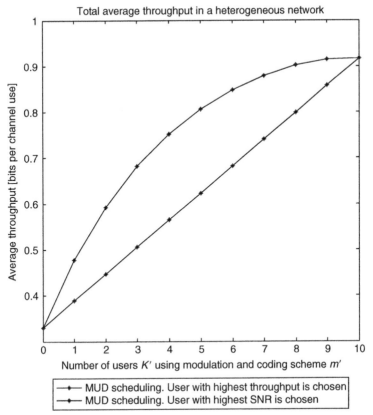

Figure 4.4 *Theoretical total average throughput for an heterogeneous system with K + K' = 10 users. Rayleigh fading channel with average SNR = 0 dB. We used values corresponding to modulation and coding schemes m = 1 and m' = 2 in Table 2.3 in Chapter 2.*

average throughput is easily computed as

$$\bar{R}_k = R^{(m)} \int_0^\infty PSR^{(m)}(\gamma_k)(F(\gamma_k))^{K+K'-1} f(\gamma_k) d\gamma_k$$

$$= R^{(m)} \sum_{k=0}^{K+K'-1} \binom{K+K'-1}{k} (-1)^k \left(\frac{1}{k+1}\right) e^{-(k+1)\frac{\gamma^{(m)}}{\bar{\gamma}}} \quad (4.24)$$

The difference between the two curves in the figure highlights the importance of using cross-layer information when performing MUD scheduling in heterogeneous systems.

p0460   Although the previous example is a very simple example it pro-
         vides a theoretical framework to study the importance of cross-
         layer information in MUD scheduling.

p0470   Imagine a more realistic scenario where K users with legacy ter-
         minals using four modulation and coding schemes share the radio
         resources with K' users with advanced terminals using eight
         modulation and coding schemes.

p0480   Figure 4.5 shows the simulated total average throughput when
         MUD scheduling policy (4.14) is implemented in this situation.
         We used $\theta_k = \theta_k = 1$ for $k = 1, \ldots, K$ and $k' = 1, \ldots, K'$. In the

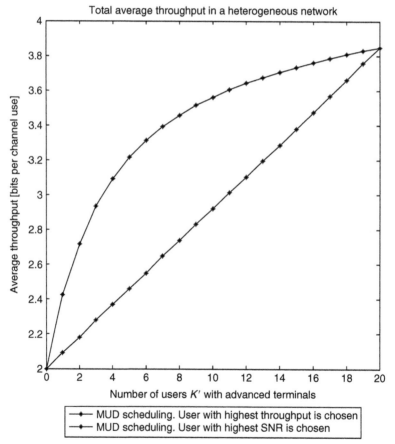

f0050   **Figure 4.5** *Total average throughput for an heterogeneous system with $K + K' = 20$
         users. Rayleigh fading channel with average SNR = 10 dB. The K users use four modu-
         lation and coding schemes according to Table 2.3 in section 2.7. The K' users use eight
         modulation and coding schemes according to Table 2.3 in section 2.7.*

figure, the number of K' users using advanced terminals changes from 0 to 20. That is, starting from a scenario with 20 users using legacy terminals (four modulation and coding schemes), the total average throughput is computed for different scenarios until all users use advanced terminals (eight modulation and coding schemes). Clearly, in the case where all users use advanced terminals, the total average throughput is maximum.

The most important conclusions are derived when the total average throughput obtained through MUD scheduling (4.14) is compared with that of a MUD policy that schedules the user with the highest SNR without taking into account any additional heterogeneity among users. Results in Figure 4.5 are clear, because $\arg \max_k \left\{ R^{(m^*)} PSR^{(m^*)}(\gamma_k) \right\} \neq \arg \max_k \left\{ \gamma_k \right\}$, the second policy is suboptimal and the total average throughput is maximized by using cross-layer information at the scheduler. We observe that the difference between the two policies is maximal when the heterogeneity among users is also maximal. That is, around the point where 10 users use legacy terminals and 10 users use advanced terminals.

Note that the MUD policy that schedules the user with the highest SNR is fair in the sense that, on average, all users will be scheduled the same number of times (this is true because all users experience the same channel conditions). However, the cross-layer MUD scheduling policy (4.14) favors users with advanced terminals because, on average, these are the ones that better exploit the channel. In this case, such potential unfairness of the cross-layer MUD scheduling policy can be compensated (if desired by the operator) by allocating more priority to legacy users.

4.3.1 HMUD and feedback information

Following the example in the previous section, in the uplink, advanced terminals with eight modulation and coding schemes require at least three bits of feedback information so the AP can feed back the optimal modulation and coding scheme for transmission. However, legacy terminals with four modulation and coding schemes require only two feedback bits. Then, another interesting point is observed: what are the effects on the total average throughput if the amount of feedback used by advanced users is reduced? To this aim, assume that users with advanced terminals decide to use only the four highest (less robust) modulation and coding schemes. In that way, because advanced terminals only use four modulation and coding schemes, only two feedback bits are necessary. Now, the scenario is composed of K users with legacy terminals that transmit data using the more robust modulation and coding schemes (from $m = 1$ to $m = 4$) and K' users with advanced terminals that transmit data using the less robust modulation and coding schemes (from $m' = 5$ to $m' = 8$). A comparison between a scenario with full feedback and a scenario with reduced feedback is given in Figure 4.6.

p0520   The total average throughput is presented in Figure 4.7 and Figure 4.8. From Figure 4.7 we observe that the higher the number of terminals in the network, the fewer the differences between a system with full feedback information where the advanced terminals use eight modulation and coding schemes with a system with reduced feedback information where the advanced terminals decide to reduce feedback by using only the highest (less robust) four modulation and coding schemes. The reason for this is that when the number of terminals is high, the selected terminal (which is the best one among all) experiences very good channel conditions and, hence, always uses high modulation and coding schemes. When the number of terminals is low, it might happen

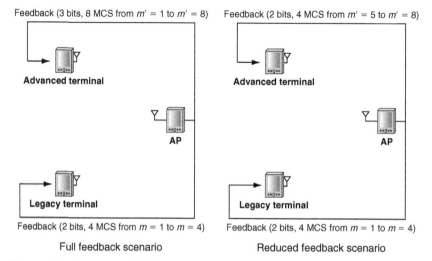

f0060   **Figure 4.6** *Scenario with full feedback vs scenario with reduced feedback.*

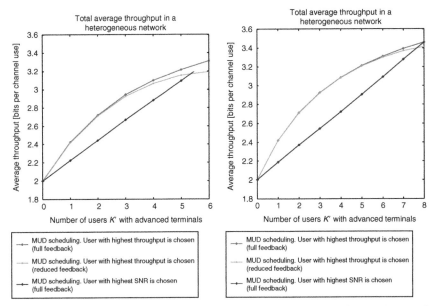

Figure 4.7 *Total average throughput for a system with K + K' = 6 (left) and*   *f0070*
K + K' = 8 (right). Rayleigh fading channel with average SNR = 10 dB.

that the SNR of the best terminal is not so good and, therefore, the optimal modulation and coding scheme is one of the lowest. If the system is heterogeneous, there is a legacy terminal that can exploit the low SNR (because it uses low modulation and coding schemes). However, if the system is composed of only advanced terminals, none of them can exploit the low SNR unless they switch to use full feedback information and, hence, eight modulation and coding schemes are used instead of four. This is the reason why, at some point, the total average throughput of a system with the advanced terminals using only four modulation and coding schemes is less than that of the system with advanced terminals using the eight modulation and coding schemes. Similar conclusions are obtained from Figure 4.8 where variations on the average SNR are considered instead of variations on the total number of users.

0530 In conclusion, by providing the AP with the proper system information, the amount of feedback information can be reduced in some heterogeneous scenarios without losing system performance.

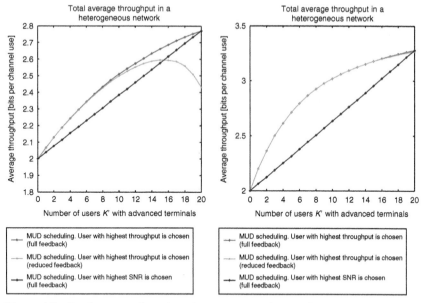

Figure 4.8 *Total average throughput for a system with K + K' = 20. Rayleigh fading channel with average SNR = 6 dB (left) and average SNR = 8 dB (right).*

s0100

4.4 Summary

p0540

In this chapter, a cross-layer approach to the MUD scheduling was presented. In particular, the general power allocation problem was reduced to an MUD scheduling problem where, every time instant, the best user is scheduled. The cross-layer perspective is taken by defining the best user in terms of throughput instead of information-theory rate or SNR. The average throughput region of MUD scheduling was studied and compared with the average information-theory rate region.

p0550

Special emphasis was put in systems where HMUD can be exploited. In this case, a closed form approximation for the average throughput was presented and it was shown that, at some levels of HMUD, the amount of feedback information can be reduced without reducing the total average throughput. Furthermore, the MUD scheduling policy that selects the user with maximum throughput is always compared with the MUD policy that selects the user with maximum SNR. This comparison shows the reader of graphical intuition the importance of cross-layer information.

References

[1]   R. Knopp and P.A. Humlet, Information capacity and power control in single-cell multiuser communications, *ICC, 1995*, June 1995.

[2]   D. Gesbert and M.-S. Alouini, Selective multi-user diversity, *ISSPIT*, 2003.

[3]   F. Florén, O. Edfors and B.-A. Molin, Throughput analysis of three multiuser diversity schemes, *Proc. Vehicular Technology and Communications (VTC)*, April 2003.

[4]   C. Mun, et al., Exact capacity analysis of multiuser diversity combined with transmit diversity, *IEE Electronics Letters*, Vol. 40, Issue No. 22, Oct. 2004.

[5]   P. Viswanath, et al., Opportunistic beamforming using dumb antennas, *IEEE Trans. Information Theory*, Vol. 48, Issue No. 6, June 2002.

[6]   S. Boyd and L. Vandenberghe, *Convex Optimization*, Cambridge University Press, 2004.

[7]   D. Tse and P. Viswanath, *Fundamentals of Wireless Communication*, Cambridge University Press, May 2005.

[8]   D. Tse and S. Hanly, Multi-access fading channels: Part I: Polymatroid structure, optimal resource allocation and throughput capacities, *IEEE Trans. Information Theory*, Vol. 44, Issue No. 7, pp. 2796–2815, Nov. 1998.

[9]   L. Li and A. Goldsmith, Capacity and optimal resource allocation for fading broadcast channels: Part I: Ergodic capacity, *IEEE Trans. Information Theory*, Vol. 47, Issue No. 3, pp. 1083–1102, March 2001.

5

0005

Cross-layer resource allocation in SIMO systems

0010

The optimal power allocation strategy that maximizes the ergodic sum capacity and the capacity region in the uplink of a multi-user MIMO wireless system has been extensively investigated in [1–3]. The general system architecture is that multiple antenna users communicate with a base station or AP provided also with multiple antennas. Based on the CSI, the resource allocation unit at the AP decides the power to be allocated to each transmitting antenna and sends such information through a signaling channel. When the feedback information is limited to one scalar the best option is to allocate power to only one transmitting antenna. Hence, the advantages of considering a multi-user SIMO system in limited feedback scenarios [4] are shown.

0020

The optimal transceiver architecture that achieves the ergodic sum capacity is rather complex and suboptimal architectures might be considered instead. In that case, the optimal power allocation that maximizes the ergodic sum of information-theory rates for a given architecture (MMSE, ZF, etc.) is to be found. However, such an optimization problem is not always convex due to the intricate relationship of the interferences [5]. Indeed, when using a MMSE receiver the problem is not convex but in the particular case of a ZF receiver, interferences are nulled at the expense of increasing the noise power and, hence, the optimization problem becomes convex. The power allocation problem is then converted to a spatial multiplexing and scheduling problem where users either transmit at maximum power or do not transmit [6]. The optimal set of users that transmit at maximum power depends on the instantaneous channel realization.

p0030 These results are generalized in this chapter where we make use of the general cross-layer resource allocation framework that has been presented all along in this book. The optimal resource allocation that maximizes the weighted sum of spectral efficiencies is shown to be also a spatial multiplexing and scheduling policy when a ZF beamformer is used. Recall that the maximization of the weighted sum of spectral efficiencies is our means of studying the average spectral efficiency region and the total average spectral efficiency (or sum of average spectral efficiencies).

p0040 By considering a ZF beamformer, the optimization problem is convex and the optimal solution relies on the use of complete CSI at the base station, which might be a computationally complex costly solution: first, because the need for instantaneous and complete CSI imposes the design of specific signaling channels in order to estimate the channel state of all users even when only a reduced set of users will be scheduled; and second, because optimal policies typically imply an exhaustive search over all the possible sets of users. Indeed, the number of possible sets is $\sum_{m=1}^{M} \binom{K}{m}$. Note that even for relatively small numbers such as $M = 4$ antennas and $K = 10$ users, this already corresponds to an exhaustive search over 385 possible sets.

p0050 On the other hand, there are the low complexity scheduling algorithms, namely the random scheduler and the round robin (RR) scheduler. Such schedulers do not use instantaneous and complete CSI to schedule users but schedule users either randomly or sequentially thus reducing the scheduling complexity. The main difference between a multi-user SISO system and a multi-user SIMO system is that in the latter case many users can be simultaneously multiplexed because of the additional spatial dimension.

p0060 In this chapter we focus also on analyzing the performance of low complexity scheduling in multi-user SIMO systems. As in previous chapters, instead of evaluating system performance in terms of the ergodic sum of information-theory rates, we take a cross-layer approach and consider the total average throughput as the utility function to be evaluated.

5.1 Cross-layer resource allocation with a ZF beamformer

Consider a multi-user SIMO centralized wireless network with K single antenna mobile users that have to be served by an AP provided with M antennas. A perfect signaling channel is considered so the AP acquires perfect knowledge of the user's channel state. According to such CSI, resources are allocated to the users through a polling mechanism. Then, a mobile user can only transmit upon a previous reception of a polling packet.

On the one hand, the MUD scheduling strategy in Chapter 4, which schedules the user with the maximum spectral efficiency, could also be considered in this multi-user SIMO scenario. Then, the system would behave as an opportunistic scheduling system with the antennas devoted to increase the user's SNR (with the consequent increase on its spectral efficiency). Such gain in the SNR is known as the spatial diversity (SD) gain. On the other hand, the additional spatial dimension could be exploited to multiplex multiple simultaneous transmissions. This strategy is commonly known as spatial division multiple access (SDMA) and its main advantage is that multiple users can transmit data simultaneously. Clearly, the more antennas used to exploit SDMA gain, the fewer antennas used to exploit the SD gain and vice versa. In that sense, the main objective of the resource allocation unit is to dynamically allocate power to users in order to obtain the optimal combination between the spatial multiplexing benefits provided by the SDMA gain and the scheduling benefits provided by the SD gain.

In this section, we show how the general power allocation problem is reduced to a spatial multiplexing and scheduling optimization problem when a ZF beamformer is used. Then, an optimal set $\mathbb{K}_\theta^*(\mathbf{H})$ of users, with all users simultaneously transmitting at maximum power, is selected every time instant. This set of users provides the optimal combination between SDMA and SD gains or, similarly, the optimal combination between spatial multiplexing and scheduling.

Assume that a ZF beamformer is performed at the AP so that signals from the transmitting users are independently received by forming strong beams to the desired user and nulls to the interfering users.

This is illustrated in Figure 5.1. As explained in the appendix of Chapter 3, interferences are eliminated at the expense of increasing the noise power. Summarizing, the signal processing matrix applied at the AP is $\mathbf{V} = (\mathbf{H}_{\mathbb{K}}^{H}\mathbf{H}_{\mathbb{K}})^{-1}\mathbf{H}_{\mathbb{K}}^{H}$ where $\mathbf{H}_{\mathbb{K}}$ is a matrix containing the columns of $\mathbf{H}$ corresponding to the transmitting users defined by the set $\mathbb{K}$. The SNR is shown to be

$$\gamma_k^{ZF} = \frac{p_k}{\sigma^2} \frac{1}{[(\mathbf{H}_{\mathbb{K}}^{H}\mathbf{H}_{\mathbb{K}})^{-1}]_{kk}} \tag{5.1}$$

with $[(\mathbf{H}_{\mathbb{K}}^{H}\mathbf{H}_{\mathbb{K}})^{-1}]_{kk}$ defining the kth element of the diagonal of $(\mathbf{H}_{\mathbb{K}}^{H}\mathbf{H}_{\mathbb{K}})^{-1}$. Recall that the signal processing procedure is limited to the fact that the cardinality of $\mathbb{K}$ is less than the number of antennas M. Otherwise, when $|\mathbb{K}| > M$, $\gamma_k^{ZF} = 0$.

Similarly as in Chapter 3, we define the priority vector as $\theta = \{\theta_1, \ldots, \theta_K\}$ and the resource allocation problem in the form of (3.17). Indeed, the optimal power allocation policy $\mathbf{p}_{\theta}^{*}(\mathbf{H}) = \{p_1^{*}, \ldots, p_K^{*}\}$, which maximizes the weighted sum of spectral efficiencies, is to be found given the uplink instantaneous individual power constraints $p_k \leq P \: \forall \: k$. The main interest on the optimal solution $\mathbf{p}_{\theta}^{*}(\mathbf{H})$ is because this

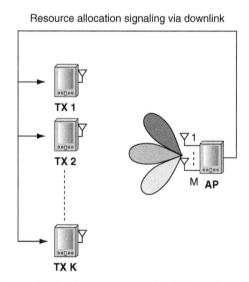

Figure 5.1   *Multi-user SIMO wireless system with a ZF beamformer.*

provides the power allocation solution that makes the system operate at any desired achievable point or, equivalently, at any point in the average spectral efficiency region.

For each channel realization, the set of transmitting users is given by $\mathbb{K} = \{k \mid k \in \{1, \ldots, K\}; p_k > 0\}$ and the weighted sum of spectral efficiencies that will be maximized can be expressed as

$$
\begin{aligned}
R_\theta(\mathbf{H}, \mathbf{p}(\mathbf{H})) &= \sum_{k=1}^{K} \theta_k R_k(\mathbf{H}, \mathbf{p}(\mathbf{H})) \\
&= \sum_{k \in \mathbb{K}} \theta_k R_k(\mathbf{H}_\mathbb{K}, \mathbf{p}(\mathbf{H}))
\end{aligned}
\tag{5.2}
$$

Clearly, the sum of spectral efficiencies is convex with

$$
\frac{\partial \sum_{k \in \mathbb{K}} \theta_k R_k(\mathbf{H}_\mathbb{K}, \mathbf{p}(\mathbf{H}))}{\partial R_k(\mathbf{H}_\mathbb{K}, \mathbf{p}(\mathbf{H}))} \geq 0 \qquad \forall R_k(\mathbf{H}_\mathbb{K}, \mathbf{p}(\mathbf{H})) \geq 0 \quad (5.3)
$$

Our aim is to show that the general power allocation problem that maximizes (5.2) can be converted into a simplified spatial multiplexing and scheduling problem where only the optimal set of simultaneously transmitting users is to be found. This can be easily demonstrated when a ZF beamformer is considered.

Let us consider a set of transmitting users $\mathbb{K}$ and a resource allocation vector $\mathbf{p}(\mathbf{H}) = \{p_1, \ldots, p_K\}$ where $p_k \neq P$ for a user $k \in \mathbb{K}$ and, by definition, $p_k = 0$ for all $k' \notin \mathbb{K}$. Consider another resource allocation vector $\mathbf{p}'(\mathbf{H}) = \{p_1, \ldots, P, \ldots, p_K\}$ where $p_k = P$ and also $p_{k'} = 0$ for all $k' \notin \mathbb{K}$. Then, assuming the use of a ZF beamformer and that the spectral efficiency is defined as $R_k(\mathbf{H}_\mathbb{K}, \mathbf{p}(\mathbf{H})) = \log_2(1 + \gamma_k^{ZF})$ or $R_k(\mathbf{H}_\mathbb{K}, \mathbf{p}(\mathbf{H})) = R^{(m^*)} PSR^{(m^*)}(\gamma_k^{ZF})$, it is true that $R_k(\mathbf{H}_\mathbb{K}, \mathbf{p}(\mathbf{H})) < R_k(\mathbf{H}_\mathbb{K}, \mathbf{p}'(\mathbf{H}))$. Since the weighted sum of spectral efficiencies is convex, as given by (5.3), the inequality $\sum_{k \in \mathbb{K}} \theta_k R_k(\mathbf{H}_\mathbb{K}, \mathbf{p}(\mathbf{H})) \leq \sum_{k \in \mathbb{K}} \theta_k R_k(\mathbf{H}_\mathbb{K}, \mathbf{p}'(\mathbf{H}))$ holds. In consequence, we conclude that given any set of transmitting users $\mathbb{K}$, the optimal power allocation policy is such that $p_k = P$. That is, all users in the set transmit at their maximum individual powers. However, the problem now is to find the optimal set of transmitting users that maximizes the weighted sum of spectral efficiencies.

p0160   In summary, the general resource allocation problem

$$\mathbf{p}_\theta^*(\mathbf{H}) = \arg\max_{\mathbf{p}(\mathbf{H})} \left(\sum_{k=1}^{K} \theta_k R_k(\mathbf{H}, \mathbf{p}(\mathbf{H})) \right) \qquad (5.4)$$

with instantaneous individual power constraints has been converted to a resource allocation problem where $\mathbf{p}(\mathbf{H}) = \{p_1, \ldots, p_K\}$ is set to $p_k = P$ for all $k \notin \mathbb{K}$ and $p'_k = 0$ for all $k' \notin \mathbb{K}$ and where an optimal set of transmitting users $\mathbb{K}_\theta^*(\mathbf{H})$ is to be found such that

$$\mathbb{K}_\theta^*(\mathbf{H}) = \arg\max_{\mathbb{K}} \left(\sum_{k\in\mathbb{K}} \theta_k R_k(\mathbf{H}_\mathbb{K}, \mathbf{p}(\mathbf{H})) \right) \qquad (5.5)$$

p0170   Then, depending on the channel state $\mathbf{H}$, the optimal set of transmitting users is to be found. Note that the policy defined by (5.5) combines spatial multiplexing and scheduling. On the one hand, the cardinality of the optimal set of transmitting users might be greater than one, indicating that spatial multiplexing is performed. On the other hand, the fact that an optimal set of users is chosen among all the users in the network indicates that, for given channel conditions, some users might be better than others. Hence, MUD scheduling is exploited. Note also that when the number of receiving antennas is equal to one, i.e. $M = 1$, this policy is reduced to the MUD scheduling policy presented in Chapter 4.

p0180   The optimal solution (5.5) must be found through an exhaustive search over all the $\sum_{m=1}^{M} \binom{K}{m}$ possible sets. Even for moderate K and M, the total search space might become very large. Consequently, in order to provide the reader with a practical approach to the resource allocation problem, low complex resource allocation strategies are presented later in this chapter. However, with the aim of providing a deeper understanding of cross-layer resource allocation in SIMO systems, we will study the average throughput region first.

s0020
5.2 The average throughput region of spatial multiplexing and scheduling

p0190   The objective of maximizing the weighted sum of spectral efficiencies is because by governing the vector of relative priorities

$\theta = \{\theta_1, \ldots, \theta_K\}$, the designer can make the system work at any desired achievable point. The set of achievable points is the average spectral efficiency region. When a cross-layer approach is taken the spectral efficiency figure of merit is the average throughput and, with a ZF beamformer at the receiver, the average throughput region is formulated as

$$\Omega = \left\{ \sum_{k=1}^{K} \theta_k \bar{R}_k \mid \theta_k \leq 1, \sum_{k=1}^{K} \theta_k = 1, \bar{R}_k \in S, k \in \{1, \ldots, K\} \right\} \quad (5.6)$$

where

$$S = \bigcup_{\mathbf{P} \in \mathbb{P}} \left\{ \bar{\mathbf{R}} : \bar{R}_k = E\{R_k(\mathbf{H}, \mathbf{p}(\mathbf{H}))\}, k \in \{1, \ldots, K\} \right\} \quad (5.7)$$

and

$$R_k(\mathbf{H}, \mathbf{p}(\mathbf{H})) = R^{(m^*)} PSR^{(m^*)}(\gamma_k^{ZF}) \quad (5.8)$$

In principle, the points at the boundary of Ω are obtained by governing the priority vector θ in (5.4). However, because of the equivalence between (5.4) and (5.5), the boundary of the average throughput region can also be obtained through the variable θ in (5.5). By means of the priority vector θ, the resource allocation unit decides the individual average throughput allocated to each user. Indeed, from (5.1) and (5.5), the individual average throughput for a given vector θ is

$$\bar{R}_k = E_{\mathbf{H}} \left\{ \sum_{\substack{\mathbb{K}_\theta^*(\mathbf{H}) \\ \text{s.t. } k \in \mathbb{K}_\theta^*(\mathbf{H})}} R^{(m^*)} PSR^{(m^*)}(\gamma_k^{ZF}) \right\}$$

$$= E_{\mathbf{H}} \left\{ 1\{\mathbb{K}_\theta^*(\mathbf{H}) = \{k\}\} R^{(m^*)} PSR^{(m^*)}(\gamma_k^{ZF}) \right\} \quad (5.9)$$

$$+ E_{\mathbf{H}} \left\{ \sum_{\substack{\mathbb{K}_\theta^*(\mathbf{H}) \neq \{k\} \\ \text{s.t. } k \in \mathbb{K}_\theta^*(\mathbf{H})}} R^{(m^*)} PSR^{(m^*)}(\gamma_k^{ZF}) \right\}$$

where $1\{\mathbb{K}_\theta^*(\mathbf{H}) = \{k\}\}$ is the indicator function taking value 1 when $\mathbb{K}_\theta^*(\mathbf{H}) = \{k\}$ and 0 otherwise. The first sum in (5.9) corresponds to the individual average throughput when terminal k is scheduled

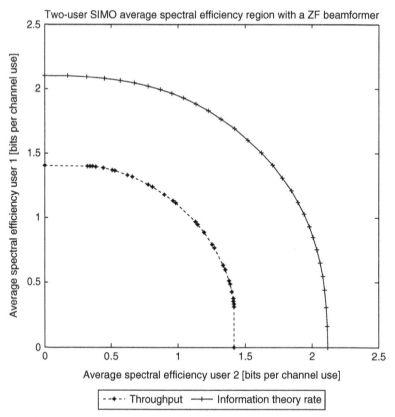

Figure 5.2 *Two-user SIMO average spectral efficiency region with a ZF beamformer. Throughput is computed using the eight modulation and coding schemes with packet length equal to 1152 information bits, as in Table 2.3 in section 2.7. A Rayleigh fading channel with average SNR = 3 dB is considered. The number of antennas at the AP is two.*

alone and the second sum in (5.9) corresponds to the individual average throughput when terminal k is scheduled in the presence of some interference. Then, the first sum in (5.9) is the individual average throughput achieved when the SD gain is fully exploited and the second sum in (5.9) includes the gain introduced due to spatial multiplexing (or SDMA).

p0210 Obviously, if (5.8) is substituted by $R_k(\mathbf{H}_\mathbb{K}, \mathbf{p}(\mathbf{H})) = \log_2(1 + \gamma_k^{ZF})$, the average information-theory rate region is obtained instead. Examples of the average information-theory rate and the average throughput regions of a two-user SIMO system with a ZF beamformer at the AP are shown in Figure 5.2 and Figure 5.3. In

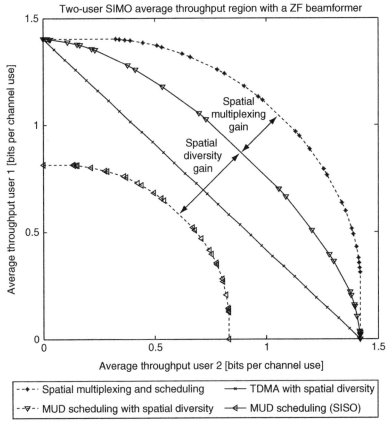

Figure 5.3 *Two-user SIMO average throughput region with a ZF beamformer and for different resource allocation strategies. Throughput is computed using the eight modulation and coding schemes with packet length equal to 1152 information bits, as in Table 2.3 in section 2.7. A Rayleigh fading channel with average SNR = 3 dB is considered and the number of antennas at the AP is two.*

f0030

Figure 5.2 a comparison between the average throughput region and the average information-theory rate region is presented. According to Shannon, if packets were infinitely long, there should exist an optimal channel coding scheme that could guarantee error-free transmission with an effective rate equal to $\log_2(1 + \gamma_k^{ZF})$. However, packets cannot be infinitely long and, hence, Shannon's limit cannot be achieved in practice. Therefore, there is always a packet error rate. Such packet error rate is considered in the throughput measure but is not considered in the information-theory rate measure. As a consequence, the gap between the two curves in Figure 5.2 appears.

p0220   In Figure 5.3, the average throughput region for different resource allocation policies is shown. Clearly, the optimal policy is given by (5.5) where for every time instant the optimal combination between spatial multiplexing and scheduling is obtained. If instead we limit ourselves to use the MUD scheduling policy presented in Chapter 4 that schedules only one user at a time, we observe that whereas there is a spatial diversity gain between SIMO and SISO systems, there is also a limitation because spatial multiplexing is not exploited. In the figure we also observe that a simple TDMA strategy (implemented with RR or random schedulers) that fully exploits spatial diversity achieves a higher average throughput region than an MUD scheduling strategy in an SISO system. Therefore, the advantage of using multiple antennas is also shown.

s0030   ## 5.3   The total average throughput of spatial multiplexing and scheduling

p0230   Previously, the average throughput region was studied by maximizing the weighted sum of throughputs. Then, the system can work at any point on the boundary of the average throughput region by modifying the vector of relative priorities θ. Moreover, each user experiences an individual average throughput (5.9) that depends on that vector of relative priorities. Recall that the sum of all individual average throughputs is called the total average throughput and that the total average throughput is maximized when all users have the same priority. In Chapter 4, the total average throughput in systems with HMUD was studied. In this section we will study the total average throughput of spatial multiplexing and scheduling.

p0240   The spatial multiplexing and scheduling policy described by (5.5) is a MUD policy that optimally combines SD and MUX gains. That is, every time the channel conditions change, the policy optimally combines the fact that some terminals may have better channel conditions than others (MUD policy) with the fact that multiple antennas designated to one terminal increase the link reliability (SD gain) and that multiple antennas designated to multiple terminals increase the number of terminals simultaneously transmitting data (MUX gain). Since the number of spatially multiplexed users changes from time to time, the average set cardinality given by $\bar{K} = E\left\{\left|\mathbb{K}_{\theta}^{*}(\mathbf{H})\right|\right\}$, i.e. the average number of spatially multiplexed users, is a good measure to understand the compromise between SD and MUX gains. Therefore, when system conditions are favoring MUX gain, $\bar{K}$ is high. Conversely, when system conditions are favoring SD gain, $\bar{K}$ is low.

In order to evaluate how different system conditions influence the decisions taken by the policy (5.5), Figure 5.4 shows the average set cardinality $\bar{K}$ for a different number of antennas at the AP and for a different number of terminals in the network. In this figure, we assume

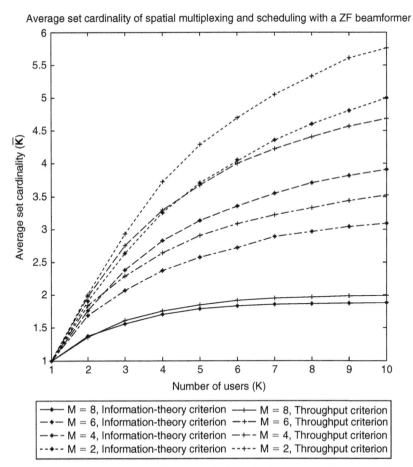

Figure 5.4 *Average set cardinality of spatially multiplexed users. Throughput is computed using the eight modulation and coding schemes with packet length equal to 1152 information bits, as in Table 2.3 in section 2.7. Rayleigh fading channel with average SNR = −6 dB.*

all users with equal priority. Besides, as usually done in this book, results obtained when $R_k(\mathbf{H}, \mathbf{p}(\mathbf{H})) = R^{(m*)} PSR^{(m*)} (\gamma_k^{ZF})$ are compared with results obtained when $R_k(\mathbf{H}, \mathbf{p}(\mathbf{H})) = \log_2(1 + \gamma_k^{ZF})$. In general, the average set cardinality increases with the number of users. This is an expected result because as the number of users increases the probability to find a high number of non-interfering users also increases because of the MUD. Ideally, for a very large number of users compared to the number of antennas, the optimal policy (5.5) should always find a set of users with cardinality equal to the number of antennas (which is the maximum set cardinality for a ZF beamformer). On the other hand, for a low number of users compared to the number of antennas, all users should be simultaneously multiplexed. In general, for a reasonable number of users and antennas we observe that, on average, the number of terminals spatially multiplexed is higher when considering an information-theory criterion than when considering a throughput criterion. The reason for this is that the sum in (5.5) depends on the number of users in $\mathbb{K}$ in two different ways. In general, the more the users in $\mathbb{K}$, the more the summands in (5.5) but at the same time, the more the users in $\mathbb{K}$, the lower the SNR and, hence, the lower the individual spectral efficiency. Therefore, the set $\mathbb{K}_\theta^*(\mathbf{H})$ is the optimal combination between these two effects. The dependence on the number of users of the sum of spectral efficiencies in (5.5) is different if we consider information-theory rate or throughput. Indeed, the curve of information-theory rate as a function of the SNR, given by $\log_2(1 + \gamma_k)$, is, in general, smoother than that of the throughput, given by $R^{(m*)}PSR^{(m*)}(\gamma_k)$. Therefore, to reduce the SD gain in favor of the MUX gain or, equivalently, to reduce the individual SNR in favor of increasing the number of users in the set is more critical in the throughput case.

p0260

From results shown in Figure 5.4, it is easy to imagine that a resource allocation unit that selects users' transmissions according to (5.5) with $R_k(\mathbf{H}, \mathbf{p}(\mathbf{H})) = \log_2(1 + \gamma_k^{ZF})$ might lead to undesirable throughput results because information-theory rates are not achievable but, instead, throughput is the practical figure of merit. This is shown in Figure 5.5. In that figure, the total average throughput for a different number of antennas and users in the system is analyzed. Two cases are considered, the total average throughput when the optimal set $\mathbb{K}_\theta^*(\mathbf{H})$ is chosen according to (5.5)

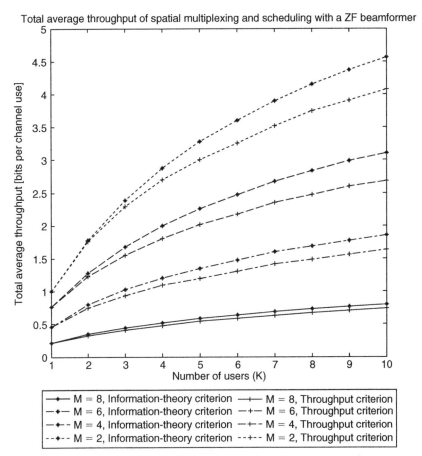

Total average throughput of spatial multiplexing and scheduling with a ZF beamformer

Figure 5.5 *Total average throughput of spatial multiplexing and scheduling.* f0050
Throughput is computed using the eight modulation and coding schemes with packet
length equal to 1152 information bits, as in Table 2.3 in section 2.7. Rayleigh fading
channel with average SNR = −6 dB.

with $R_k(\mathbf{H}, \mathbf{p}(\mathbf{H})) = \log_2(1 + \gamma_k^{ZF})$ and the total average throughput when the optimal set $\mathbb{K}_\theta^*(\mathbf{H})$ is chosen according to (5.5) with
$R_k(\mathbf{H}, \mathbf{p}(\mathbf{H})) = R^{(m*)} PSR^{(m*)}(\gamma_k^{ZF})$. Then, Figure 5.5 represents
the total average throughput when the average number of spatially
multiplexed users corresponds to that plotted in Figure 5.4. Note
that in Figure 5.5 we plot throughput and not information-theory
rate. Then, as expected, we observe that the total average throughput is higher when users are selected using a throughput criterion
rather than an information-theory rate criterion.

p0270   As explained in the previous section, the policy (5.5) that chooses the optimal set $\mathbb{K}^*_\theta(\mathbf{H})$ implies an exhaustive search over all the $\sum_{m=1}^{M} \binom{K}{m}$ possible sets of users. Since this might be a non-desired solution in terms of implementation, low complex resource allocation strategies are presented in the next section.

s0040   ## 5.4   Low complexity spatial multiplexing and scheduling policy

p0280   Typically, the low complexity scheduling algorithms are the random scheduler and the RR scheduler. Such schedulers do not use CSI to schedule terminals but schedule terminals either randomly or sequentially thus reducing the scheduling complexity. When multiple antennas are used, many terminals can be simultaneously scheduled because of the additional spatial dimension. In this section we focus on analyzing the performance of low complexity spatial multiplexing and scheduling policies.

p0290   Let us consider the SIMO multiple access channel presented in Figure 5.1 where K terminals with one antenna each communicate with an AP provided with M antennas. Previous to the transmission of packets, a set of users receives access to the channel through the feedback channel. Then, the channel of the selected users is estimated and a ZF beamformer is performed to receive the transmitted packets.

p0300   In order to reduce complexity of the spatial multiplexing and scheduling policy, we assume that decisions on what is the set of transmitting users are taken without the complete and instantaneous CSI, i.e. without the channel matrix $\mathbf{H}$. Further, for the sake of simplicity, we assume that all users have the same relative priority. Then, we denote the set of spatially multiplexed users by $\mathbb{K}$, where there is no dependence on $\mathbf{H}$ and θ. Once the set $\mathbb{K}$ of active users is chosen, the channel matrix $\mathbf{H}_\mathbb{K}$ is estimated to perform a ZF beamforming. The resulting SNR for each user k in the set $\mathbb{K}$ is, therefore, given by (5.1).

p0310   With the aim of serving all the K users in the system, the set of selected users must change from time to time. However, the dependence of $\mathbb{K}$ on the time-varying channel $\mathbf{H}$ has been eliminated for complexity reasons. Then, we define a random policy characterized by a vector $\mathbf{t} = \{t_{\mathbb{K}} : \mathbb{K} \subseteq P\{1, \ldots, \mathbb{K}\}\}$ ($P\{1, \ldots, \mathbb{K}\}$ is used to indicate the set containing all the partitions of the set $\{1, \ldots, K\}$) with $t_{\mathbb{K}} \in [0, 1]$ and $\sum_{\mathbb{K} \subseteq P\{1, \ldots, K\}} t_{\mathbb{K}} \leq 1$ where $t_{\mathbb{K}}$ is the probability that terminals in the set $\mathbb{K}$ are selected at a given time. In other words, the random policy $\mathbf{t}$ is a time-sharing policy that substitutes the time-varying decisions (5.5) by probabilistic decisions.

320 Due to the fact that the set of active users $\mathbb{K}$ does not depend on $\mathbf{H}$ but on the policy $\mathbf{t}$, the individual average throughput can be expressed as

$$\bar{R}_k = \sum_{\substack{\mathbb{K} \subseteq P\{1, \ldots, K\} \\ \text{s.t. } k \in \mathbb{K}}} t_{\mathbb{K}} E_{\mathbf{H}} \left\{ R^{(m^*)} PSR^{(m^*)}(\gamma_k^{ZF}) \right\} \quad (5.10)$$

and the total average throughput is the sum of all the individual average throughputs given by $\bar{R} = \sum_k \bar{R}_k$.

330 While the optimal set (5.5) changes according to the CSI, when no CSI is available, there is no information to decide how the set of active users should change. Then, the design of the random policy $\mathbf{t}$ accepts many different criteria depending on the designer. However, because there is no information on how to dynamically increase or reduce the set of transmitting users, a simple and reasonable criterion would be to always spatially multiplex and schedule users in sets of equal size $\mathbf{K}$. That is, to set the condition $\sum_{\substack{\mathbb{K} \subseteq P\{1, E, \ldots, K\} \\ s.t. |\mathbb{K}|=\mathbf{K}}} t_{\mathbb{K}} = 1$. In this case, the optimal set size that maximizes the total average throughput with no CSI can be obtained.

0340 For the purpose of finding the optimal set size, assume that all users are homogeneous and behave similarly on average. Then, the expectation $E_{\mathbf{H}} \left\{ R^{(m^*)} PSR^{(m^*)}(\gamma_k^{ZF}) \right\}$ is the same for any user $k \in \mathbb{K}$ and for any set $\mathbb{K}$ with cardinality $\mathbf{K} = |\mathbb{K}|$. Recall that γ_k^{ZF} depends on $\mathbb{K}$ and $\mathbf{H}$ through expression (5.1). Formally speaking, given any user $k \in \mathbb{K}$, any user $k' \in \mathbb{K}'$ and sets $\mathbb{K}$ and $\mathbb{K}'$ with cardinality $\mathbf{K} = |\mathbb{K}| = |\mathbb{K}'|$, it is true that

$E_\mathbf{H}\left\{R^{(m^*)}PSR^{(m^*)}(\gamma_k^{ZF})\right\} = E_\mathbf{H}\left\{R^{(m^*)}PSR^{(m^*)}(\gamma_{k'}^{ZF})\right\}$. Therefore, for any random policy $\mathbf{t}$ such that $\sum_{\substack{\mathbb{K} \subseteq P\{1,\,...,K\} \\ s.t.\ |\mathbb{K}|=\mathbf{K}}} t_\mathbb{K} = 1$, the total average throughput is

$$
\begin{aligned}
\bar{R}_\mathbf{K} &= \sum_{\substack{\mathbb{K} \subseteq P\{1,...,K\} \\ s.t.\ |\mathbb{K}|=\mathbf{K}}} t_\mathbb{K} \sum_{k \in \mathbb{K}} E_\mathbf{H}\left\{R^{(m^*)}PSR^{(m^*)}(\gamma_k^{ZF})\right\} \\
&= \mathbf{K}\, E_\mathbf{H}\left\{R^{(m^*)}PSR^{(m^*)}(\gamma_k^{ZF})\right\}
\end{aligned}
\tag{5.11}
$$

p0350   The index $\mathbf{K}$ is used to indicate that (5.11) is the total average throughput obtained when users are spatially multiplexed and scheduled in groups of $\mathbf{K}$ users with independence on what is the set $\mathbb{K}$ that is chosen every time, i.e. when $\sum_{\substack{\mathbb{K} \subseteq P\{1,...,K\} \\ s.t.\ |\mathbb{K}|=\mathbf{K}}} t_\mathbb{K} = 1$.

p0360   One objective of the designer would be to obtain the optimal value $\mathbf{K}^*$ such that maximizes the total average throughput given by (5.11). That is,

$$
\mathbf{K}^* = \arg\max_\mathbf{K} \bar{R}_\mathbf{K}
\tag{5.12}
$$

p0370   A special case is when $\mathbf{K}^* = 1$, that is, when $\bar{R}_1 > \bar{R}_\mathbf{K}$ for $\mathbf{K} = 2, \ldots, K$. In that case, spatial diversity is better than spatial multiplexing and, hence, the maximum total average throughput is achieved by scheduling terminals in groups of 1, i.e. following a pure TDMA strategy.

s0050   ### 5.4.1 A closed form approximation for the average throughput in Rayleigh fading channels

p0380   A closed form approximation of $\bar{R}_\mathbf{K}$ would be of great help in order to solve the optimization problem (5.12). To this end, remember that when a ZF beamformer is used, the user's SNR is given by (5.1). Then, we can define the random variable $\beta = 1/[(\mathbf{H}_\mathbb{K}^H \mathbf{H}_\mathbb{K})^{-1}]_{kk}$ and the parameter $\beta_{th}^{(m)} = \gamma_{th}^{(m)}/\bar{\gamma}$ as the SNR enhancement necessary to achieve the minimum SNR required for the modulation and coding scheme m when the average SNR is given by $\bar{\gamma} = p_k/\sigma^2$. Note that link adaptation was explained in

Chapter 2. Besides, according to the QoS section in Chapter 2, the parameter $\gamma_{th}^{(m)}$ can be seen as the minimum QoS requirement related to best-effort QoS service. Hence, the substitution of $\gamma_{th}^{(m)}$ by a parameter $\gamma_{QoS}^{(m)}$, denoting other minimum QoS requirements, should be straightforward. For given QoS requirements, the SNR enhancement parameter $\beta_{th}^{(m)}$ decreases as the average SNR increases.

When the entries of $\mathbf{H_K}$ are independent and identically distributed circular symmetric complex Gaussian random variables $CN(0,\sigma^2)$, i.e. Rayleigh fading, the random variable β is a weighted chi-square distributed variable with $r_K = 2\,(M - \mathbf{K} + 1)$ degrees of freedom and p.d.f. equal to $f_{\mathbf{K}}(\beta) = \beta^{K-\mathbf{K}}e^{-\beta}/(M - \mathbf{K})!$ [7]. Recall that the number of users in the system is denoted by K and that due to ZF restrictions the maximum set size $\mathbf{K}$ is the number of antennas M. Then, taking the expectation in (5.11), the total average throughput is computed as

$$\bar{R}_{\mathbf{K}} = \mathbf{K} \sum_{m=1}^{Mod} R^{(m)} \int_{\beta_{th}^{(m)}}^{\beta_{th}^{(m+1)}} PSR^{(m)}(\beta) f_{\mathbf{K}}(\beta) d\beta \tag{5.13}$$

where *Mod* is the total number of modulation and coding schemes and the throughput corresponding to each modulation and coding scheme m is integrated over the range where this modulation and coding scheme is optimal, according to Chapter 2. Moreover, in that chapter, an approximation of the PSR was given. This approximation can be further simplified to a step function approximation which, expressed in terms of β, is given by

$$PSR^{(m)}(\beta) \simeq \begin{cases} 0 & \text{for } \beta \leq \beta_{th}^{(m)} \\ 1 & \text{for } \beta \geq \beta_{th}^{(m)} \end{cases} \tag{5.14}$$

Then, combining (5.13) with (5.14), it is easy to obtain

$$\begin{aligned} \bar{R}_{\mathbf{K}} &= \frac{\mathbf{K}}{(M - \mathbf{K})!} \sum_{m=1}^{Mod} \left(R^{(m)} - R^{(m-1)} \right) \Gamma\left(\frac{r_{\mathbf{K}}}{2}, \beta_{th}^{(m)} \right) \\ &= \mathbf{K} \sum_{m=1}^{Mod} \left(R^{(m)} - R^{(m-1)} \right) \sum_{i=0}^{M-\mathbf{K}} \frac{(\beta_{th}^{(m)})^i e^{-\beta_{th}^{(m)}}}{i!} \end{aligned} \tag{5.15}$$

where $\Gamma(\beta,x)$ is $\int_x^\infty e^{-t}t^{\beta-1}dt$ and by definition, $R^{(m)} > R^{(m-1)}$ with $R^{(0)} = 0$.

p0410   A comparison of the total average throughput closed form approximation (5.15) with the simulated total average throughput is given in Figure 5.6. We observe that (5.15) is a fairly good approximation and the small differences are due to the step function approximation of the PSR. As noted in Chapter 2, the step

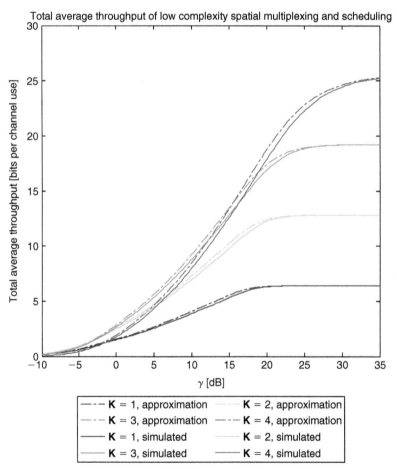

f0060   **Figure 5.6** *Total average throughput of low complexity spatial multiplexing and scheduling for different values of average SNR. A ZF beamformer is performed with M = 4 antennas. Throughput is computed using the eight modulation and coding schemes with packet length equal to 1152 information bits, as in Table 2.3 in section 2.7. A Rayleigh fading channel is considered.*

function approximation is better as the minimum QoS requirements increase. Therefore, for more stringent QoS requirements,
the approximation (5.15) would be even closer to the simulated
results. Moreover, from results in Figure 5.6 we observe that for
different ranges of the average SNR there is an optimal value of
the spatially multiplexing and scheduling set size that maximizes
throughput. The average SNR range corresponding to an optimal
value of the set size depends on the number of antennas. This is
better observed later in Figure 5.8.

Clearly, there exists a dependence of the total average throughput
on the spatially multiplexing and scheduling set size **K**. Looking
at expression (5.15), we observe that, if we increase the number
of users in the set, the throughput should increase because there
is a factor **K** multiplying. This accounts for the effect of spatial
multiplexing. However, if we increase the number of users in the
set, the throughput should decrease because the number of elements in the second summation decreases with **K**. This accounts
for the spatial diversity effect saying that the fewer the number of
users transmitting, the higher the SNR enhancement for each user.
Therefore, there exists a trade-off between spatial multiplexing
and spatial diversity with an optimal value **K*** that maximizes the
total average throughput. An example of this trade-off is shown in
Figure 5.7 where the total average throughput for different scheduling set cardinalities is shown.

Moreover, from (5.15), we observe that the more the number of
antennas, the higher the throughput. Also, the lower the $\beta_{th}^{(m)}$,
the higher the throughput or, equivalently, the higher the average
SNR, the higher the throughput. Therefore, the optimal **K***
depends on both the number of antennas M and the average SNR
$\overline{\gamma}$. Such dependence is shown in Figure 5.8 where we see that the
higher the average SNR and the number of antennas, the higher
the optimal **K***.

In this section a low complexity spatial multiplexing and scheduling policy has been studied. In fact, the policy presented is the
less complex among all the policies since it does not use any kind
of CSI. At the other extreme, there is the optimal spatial multiplexing and scheduling policy when complete CSI is available

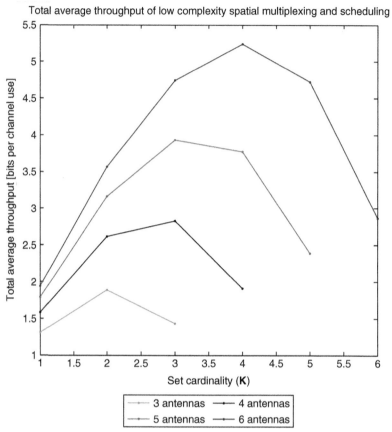

Figure 5.7 *Total average throughput of low complexity spatial multiplexing and scheduling for different scheduling set cardinalities. Throughput is computed using the eight modulation and coding schemes with packet length equal to 1152 information bits, as in Table 2.3 in section 2.7. A Rayleigh fading channel with average SNR = 0 dB is considered.*

at the resource unit. This policy was presented in previous sections. In Figure 5.9 we plot a comparison of these two policies. Note that the curve of the low complexity policy corresponds to the envelope of the curves in Figure 5.6. As expected, a throughput increase appears when the optimal policy is implemented. This throughput increase depends on the number of users in the system because with more users the MUD gain increases. However, at high SNR the optimal policy and the low complexity policy achieve similar results as both policies will always schedule a number of users equal to the number of antennas and, at the same time, the MUD gain is very low. The gap between the throughputs

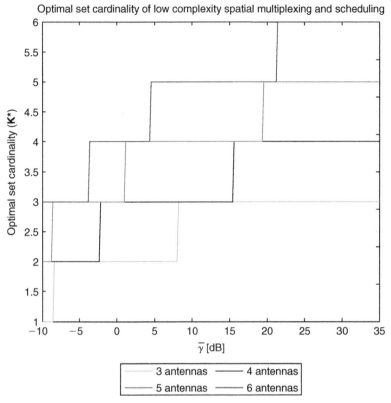

Optimal set cardinality of low complexity spatial multiplexing and scheduling

Figure 5.8 *Optimal set cardinality of low complexity spatial multiplexing and schedul-* f0080
ing for different values of average SNR. Throughput is computed using the eight modula-
tion and coding schemes with packet length equal to 1152 information bits, as in Table
2.3 in section 2.7. A Rayleigh fading channel is considered.

achieved by the two policies could be filled by investigating other suboptimal resource allocation policies such as that in [6]. Nevertheless, there will be always the compromise between complexity and optimality.

5.5 Summary

In this chapter, the topic of cross-layer resource allocation in wireless SIMO systems was covered. One of the main interesting results presented was that the optimal power allocation that maximizes the weighted sum of spectral efficiencies is equivalent to a spatial multiplexing and scheduling policy when a ZF beamformer is used. Besides, thanks to the general cross-layer resource allocation

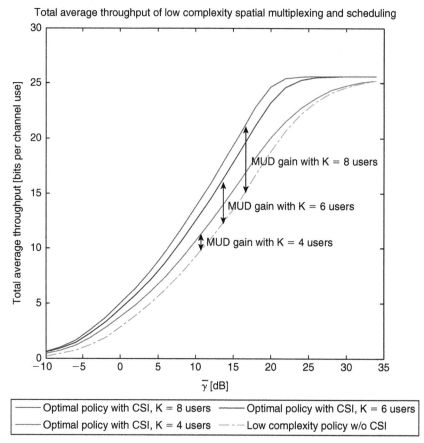

f0090

Figure 5.9   *Total average throughput of optimal and low complexity spatial multiplexing and scheduling policies for different values of average SNR. A ZF beamformer is performed with M = 4 antennas. Throughput is computed using the eight modulation and coding schemes with packet length equal to 1152 information bits, as in Table 2.3 in section 2.7. A Rayleigh fading channel is considered.*

framework presented in previous chapters, the maximization of the weighted sum of spectral efficiencies was used to study the average spectral efficiency region and the total average spectral efficiency.

p0460

However, the optimal solution was shown to rely on the use of complete CSI at the AP. Since that could be a computationally complex costly solution, a low complexity spatial multiplexing and scheduling policy was also studied in this chapter. This is presented as the less complex solution as it does not use any kind of CSI to take resource allocation decisions. Between the optimal spatial multiplexing and scheduling solution and the low complexity solution presented in this chapter there are many suboptimal alternatives that could be investigated by the research community. However, there will always be a compromise between complexity and optimality.

References

[1] W. Yu, W. Rhee and J.M. Cioffi, Optimal power control in multiple access fading channels, *Proc. IEEE ICC*, 2001.

[2] W. Yu, et al., Iterative water-filling for Gaussian vector multiple access channels, *IEEE Trans. Information Theory*, Vol. 50, pp. 145–151, Jan. 2004.

[3] W. Rhee and J.M. Cioffi, On the capacity of multiuser wireless systems with multiple antennas, *IEEE Trans. Information Theory*, Vol. 49, pp. 2580–2595, Oct. 2003.

[4] V.K.N. Lau, Y. Liu and T.-A. Chen, Optimal space-time scheduling for block fading channels with partial power feedbacks, *Bell Labs Technical Journal*, Vol. 7, Issue 3, pp. 27–46, 2003.

[5] M. Schubert, H. Boche and S. Stanczak, Strict convexity of the QoS feasible region for log-convex interference functions, *Proc. ACSSC'06*.

[6] V.K.N. Lau and Y.-K. Kwok, Performance analysis of SIMO space-time scheduling with convex utility function: zero-forcing linear processing, *IEEE Trans. Vehicular Technology*, Vol. 53, Issue No. 2, pp. 339–350, March 2004.

[7] D.A. Gore, R.W. Heath Jr. and A.J. Paulraj, Transmit selection in spatial multiplexing systems, *IEEE Communication Letters*, Vol. 6, Issue No. 11, pp. 491–493, Nov. 2002.

6

Cross-layer resource allocation in MISO systems

The general architecture of a multi-user MIMO broadcast system is composed of an AP provided with multiple antennas that communicates with multiple antenna users. Prior to data transmission, the AP acquires the CSI of the channel between the AP and the users through the feedback channel. Then, the resource allocation unit decides the power to be allocated at each transmitting antenna for transmitting data to each user. The optimal power allocation strategy that maximizes the sum capacity and the capacity region in the broadcast channel (downlink) of a multi-user MIMO wireless system has been a popular topic of research over the past few years [1]. Now, it is widely known that the capacity achieving transceiver architecture is based on DPC [2, 3]. However, DPC is rather complex and difficult to implement in practical systems because of the process of successive encoding and decoding.

Beamforming strategies can be considered as reduced complexity alternatives. Recalling the signal model presented in section 3.3 in Chapter 3, each user symbol s_k is processed by a vector $\mathbf{v}_k$ before being transmitted through the channel. The processing vector $\mathbf{v}_k$ is called the beamforming vector and represents the weight given to symbol s_k at each of the transmitting antennas. Unfortunately, not only is the design of the optimal beamformer a difficult non-convex optimization problem but also the optimal power allocation that maximizes the weighted sum of spectral efficiencies for a particular set of beamforming vectors is usually a non-convex optimization problem. Recall that the maximization of the weighted sum of spectral efficiencies is the mechanism to work on the points at the boundary of the average spectral efficiency region.

p0030 Nevertheless, a simple suboptimal beamforming strategy is the ZF beamformer. Despite its suboptimality, it is shown that for a sufficiently high number of users, the sum of information-theory rates approaches the sum capacity with DPC [4]. The main reason for this is that the ZF beamformer is combined with an exhaustive search among all possible user sets so an optimal set of users who simultaneously transmit can be found. Therefore, when the number of users is large, it is highly probable to find a set of almost non-interfering users. Another suboptimal alternative that also approaches optimal performance for a large number of users is orthogonal random beamforming. In this case, the reasoning is contradictory. Multiple orthogonal (non-interfering) random beamforming vectors are generated so that, when the number of users is large, an optimal set of users with a channel that matches the randomly generated beamformer should exist. Therefore, orthogonal random beamforming approaches the sum capacity but, compared to the ZF beamformer, its convergence with the number of users is slower. Nevertheless, with orthogonal random beamforming the amount of feedback is reduced because users do not need to feed back the full CSI, only the SNR.

p0040 Similarly, as in Chapter 5, the ZF beamformer will be considered in this chapter. Furthermore, as one of the main drawbacks of the ZF beamformer is the amount of feedback information because full CSI is needed, suboptimal ZF beamforming strategies will also be investigated in this chapter. For the sake of simplicity, we will consider resource allocation in multi-user MISO systems with single antenna terminals. An example is given in Figure 6.1. In particular, the average spectral efficiency region will be investigated after the maximization of the weighed sum of spectral efficiencies. We will observe that a water-filling power solution maximizes the weighed sum of spectral efficiencies. Moreover, it will be shown that differences between the optimal power allocation and the uniform power allocation are very low. Therefore, for the sake of simplicity, uniform power allocation will be considered so the general power allocation problem will be converted into a spatial multiplexing and scheduling problem where an optimal set of users who transmit simultaneously is to be found.

6.1 Cross-layer resource allocation with a ZF beamformer

Consider a multi-user MISO centralized wireless network with K single antenna mobile users that have to be served by an AP provided with M antennas. A perfect feedback channel is considered so the AP acquires perfect knowledge of the user's channel state. According to such CSI, power is allocated in order to transmit data to users. An example of such a network was given in Figure 3.1 in Chapter 3.

Since the multiple antennas can be used to exploit either spatial diversity gain or spatial multiplexing gain, the main objective of the resource allocation unit is to dynamically allocate power in order to obtain the optimal combination between multiplexing data to multiple users and using multiple antennas for transmitting data to a single user. Note that, as we deal with wireless channels, the MUD gain is intrinsic in multi-user systems.

In this section, we follow a divide and conquer approach and show the general spectral efficiency maximization problem as the composition of two optimization subproblems. First, the optimal power allocation for a given set of users is found. And then the optimal set of users to whom power is distributed is searched.

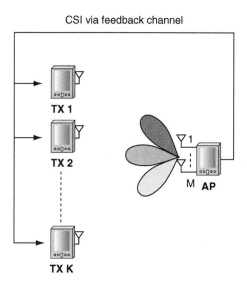

Figure 6.1 *Multi-user MISO wireless system with a ZF beamformer.*

p0080   Following the same methodology as in previous chapters, we denote the $M \times K$ SIMO channel matrix by $\mathbf{H}$, the power allocation vector by $\mathbf{p}(\mathbf{H}) = \{p_1, \ldots, p_K\}$ and the priority vector by $\boldsymbol{\theta} = \{\theta_1, \ldots, \theta_K\}$. Then, the resource allocation problem is in the form of (3.17) and the optimal power allocation policy $\mathbf{p}_{\boldsymbol{\theta}}^*(\mathbf{H}) = \{p_1^*, \ldots, p_K^*\}$, which maximizes the weighted sum of spectral efficiencies, is to be found given the downlink instantaneous total power constraint $\sum_{k=1}^{K} p_k \leq P_{total}$. Remember that, by means of $\mathbf{p}_{\boldsymbol{\theta}}^*(\mathbf{H})$, the designer can make the system operate at any desired point in the average spectral efficiency region.

p0090   The general resource allocation problem is

$$\mathbf{p}_{\boldsymbol{\theta}}^*(\mathbf{H}) = \arg \max_{\mathbf{p}(\mathbf{H})} \sum_{k=1}^{K} \theta_k R_k(\mathbf{H}, \mathbf{p}(\mathbf{H})) \qquad (6.1)$$
$$s.t. \sum_{k=1}^{K} p_k \leq P_{total}$$

Due to the total power restrictions, (6.1) is equivalent to

$$\left(\mathbb{K}_{\boldsymbol{\theta}}^*(\mathbf{H}), \mathbf{p}_{\boldsymbol{\theta}}^*(\mathbf{H}_{\mathbb{K}}) \right) = \arg \max_{\mathbb{K}, \mathbf{p}(\mathbf{H}_{\mathbb{K}})} \sum_{k \in \mathbb{K}} \theta_k R_k(\mathbf{H}_{\mathbb{K}}, \mathbf{p}(\mathbf{H}_{\mathbb{K}})) \quad (6.2)$$
$$s.t. \sum_{k \in \mathbb{K}} p_k \leq P_{total}$$

where the set $\mathbb{K}$ is a subset of users to whom the total power is distributed and $\mathbf{H}_{\mathbb{K}}$ is the channel matrix corresponding to such set of users.

p0100   From (6.2) it is clear that our optimization problem can be solved in two different stages. One stage is devoted to finding the optimal power allocation $\mathbf{p}_{\boldsymbol{\theta}}^*(\mathbf{H}_{\mathbb{K}})$ that maximizes the weighted sum of spectral efficiencies of users in a set $\mathbb{K}$. This is formulated as

$$\mathbf{p}_{\boldsymbol{\theta}}^*(\mathbf{H}_{\mathbb{K}}) = \arg \max_{\mathbf{p}(\mathbf{H}_{\mathbb{K}})} \sum_{k \in \mathbb{K}} \theta_k R_k(\mathbf{H}_{\mathbb{K}}, \mathbf{p}(\mathbf{H}_{\mathbb{K}})) \qquad (6.3)$$
$$s.t. \sum_{k \in \mathbb{K}} p_k \leq P_{total}$$

p0110   And the other stage is charged with finding the optimal set of users $\mathbb{K}_{\boldsymbol{\theta}}^*(\mathbf{H})$ over all the possible user sets. This can be expressed as

$$\mathbb{K}_{\theta}^{*}(\mathbf{H}) = \arg \max_{\mathbb{K}} \sum_{k \in \mathbb{K}} \theta_k R_k (\mathbf{H}_{\mathbb{K}}, \mathbf{p}(\mathbf{H}_{\mathbb{K}})) \tag{6.4}$$

The main advantage of such a *divide and conquer* approach is that whereas the general power allocation problem (6.1) is typically non-convex, the power allocation problem in (6.3) is convex when a ZF beamformer is considered. The main drawback is that the solution of (6.4) is based on an exhaustive search among all the possible user sets.

6.1.1 The optimal power allocation $\mathbf{p}_{\theta}^{*}(\mathbf{H}_{\mathbb{K}})$ with a ZF beamformer

For the sake of problem convexity, we assume that a ZF beamformer is performed at the AP so that users' data are independently processed by each user's ZF beamforming vector before being transmitted. This is illustrated in Figure 6.1. In particular, given a set of users $\mathbb{K}$, the signal processing matrix $\mathbf{V} = \left(\mathbf{H}_{\mathbb{K}}^{H} \mathbf{H}_{\mathbb{K}} \right)^{-1} \mathbf{H}_{\mathbb{K}}^{H}$ is applied at the AP where each column $\mathbf{v}_k$ of the matrix $\mathbf{V}$ is the *k*th user processing vector with $\left\| \mathbf{v}_k \right\|^2 = \left[\left(\mathbf{H}_{\mathbb{K}}^{H} \mathbf{H}_{\mathbb{K}} \right)^{-1} \right]_{kk}$.

Besides, as already mentioned in previous chapters, the ZF beamformer eliminates interferences at the cost of increasing the noise power by a factor $\left\| \mathbf{v}_k \right\|^2$. Therefore, the SNR is shown to be

$$\gamma_k^{ZF} = \frac{p_k}{\left\| \mathbf{v}_k \right\|^2 \sigma^2} \tag{6.5}$$

This expression is equivalent to those presented in Chapters 3 and 5. However, in the MISO broadcast channel the ZF beamformer is conformed at the transmitter side and it is worth noting that the power allocated to user k is given by

$$p_k = E \left\{ s_k s_k^* \right\} \left\| \mathbf{v}_k \right\|^2 \tag{6.6}$$

Therefore, the power restriction $\sum_{k \in \mathbb{K}} p_k \leq P_{total}$ is in fact $\sum_{k \in \mathbb{K}} \left\| \mathbf{v}_k \right\|^2 \tilde{p}_k \leq P_{total}$, where $\tilde{p}_k = E \left\{ s_k s_k^* \right\}$ is the parameter of design because $\| \mathbf{v}_k \|^2$ is determined by the channel matrix $\mathbf{H}_{\mathbb{K}}$.

p0170   Now, consider the case where the spectral efficiency $R_k(\mathbf{H}\mathbb{K}, \mathbf{p}(\mathbf{H}))$ is given by the information-theory rate $\log_2(1 + \gamma_k^{ZF})$. Then, the maximization problem (6.3) is solved by means of convex optimization method [6]. In particular, the Lagrange dual function of the maximization problem is given by

$$
\mathcal{L}\left(\mathbf{p}(\mathbf{H}_{\mathbb{K}}), \lambda\right) = -\sum_{k \in \mathbb{K}} \theta_k \log_2\left(1 + \gamma_k^{ZF}\right) + \lambda\left(\sum_{k \in \mathbb{K}} \|\mathbf{v}_k\|^2 \tilde{p}_k - P_{total}\right) \\
- \sum_{k \in \mathbb{K}} \mu_k \tilde{p}_k \tag{6.7}
$$

where λ is the Lagrange multiplier for the total power restriction and μ_k are the Lagrange multipliers for the intrinsic individual power restrictions $\tilde{p}_k \geq 0$. Using (6.5), the first derivative of the Lagrangian with respect to $\tilde{p}_k$ is

$$
\frac{\partial \mathcal{L}\left(\mathbf{p}(\mathbf{h}), \lambda\right)}{\partial \tilde{p}_k} = -\frac{\theta_k}{\sigma^2 + \tilde{p}_k} \log_2 e + \lambda \|\mathbf{v}_k\|^2 - \mu_k \tag{6.8}
$$

p0180   Therefore, we have the following sufficient and necessary KKT conditions

$$
\frac{\theta_k \log_2 e}{\sigma^2 + \tilde{p}_k} = \lambda \|\mathbf{v}_k\|^2 - \mu_k \quad 1 \leq k \leq K
$$

$$
\tilde{p}_k \geq 0 \quad 1 \leq k \leq K
$$

$$
\mu_k \geq 0 \quad 1 \leq k \leq K
$$

$$
\mu_k \tilde{p}_k = 0 \quad 1 \leq k \leq K \tag{6.9}
$$

$$
\sum_{k \in \mathbb{K}} \|\mathbf{v}_k\|^2 \tilde{p}_k - P_{total} \leq 0
$$

$$
\lambda \geq 0
$$

$$
\lambda\left(\sum_{k \in \mathbb{K}} \|\mathbf{v}_k\|^2 \tilde{p}_k - P_{total}\right) = 0
$$

From the first four conditions we obtain the following water-filling solution

$$
\begin{aligned}
p_k^* &= \left\| \mathbf{v}_k \right\|^2 \tilde{p}_k^* \\
&= \left[\frac{\theta_k \log_2 e}{\lambda} - \left\| \mathbf{v}_k \right\|^2 \sigma^2 \right]_+
\end{aligned}
\tag{6.10}
$$

where $[x]_+$ denotes $\max(0, x)$ and the water level parameter λ must be designed such that the total power constraint is satisfied.

Note that in order to solve the resource allocation problem (6.3) by means of the Lagrangian duality and the KKT conditions, the spectral efficiency $R_k(\mathbf{H}_{\mathbb{K}}, \mathbf{p}(\mathbf{H})_{\mathbb{K}}))$ must be differentiable with respect to $\tilde{p}_k$. Unfortunately, looking at the plots in Chapter 2, we observe that throughput is not a differentiable function with respect to $\tilde{p}_k$, especially when link adaptation is considered with multiple modulation and coding schemes. However, the analytical approximation of throughput envelope (2.14) can be used as an alternative. In that case, and assuming the ZF beamformer, the spectral efficiency is approximated by $R_k(\mathbf{H}_{\mathbb{K}}, \mathbf{p}(\mathbf{H}_{\mathbb{K}})) \approx \alpha_1 \log_2(1 + \alpha_2 \gamma_k^{ZF})$ where parameters α_1 and α_2 must be adjusted as shown in Chapter 2.

Therefore, repeating the Lagrangian methodology for such throughput approximation, the optimal power allocation is given by

$$
\begin{aligned}
p_k^* &= \left\| \mathbf{v}_k \right\|^2 \tilde{p}_k^* \\
&= \left(\frac{\alpha_1 \theta_k \log_2 e}{\lambda} - \frac{\left\| \mathbf{v}_k \right\|^2 \sigma^2}{\alpha_2} \right)^+
\end{aligned}
\tag{6.11}
$$

where λ is such that the total power constraint is satisfied. An efficient algorithm to determine the water level parameter λ and the power levels P_k^* is given in the annex.

A suboptimal, but simpler, power allocation strategy would be the uniform power allocation defined by

$$
p_k^{uniform} = \frac{P_{total}}{|\mathbb{K}|}
\tag{6.12}
$$

p0230   where $|\mathbb{K}|$ is the cardinality of the set $\mathbb{K}$. Differences on performance between these two strategies are shown in section 6.1.3.

s0030

6.1.2 The optimal spatial multiplexing and scheduling set $\mathbb{K}_\theta^*(\mathbf{H})$ with a ZF beamformer

p0240   The optimal power allocation $\mathbf{p}_\theta^*(\mathbf{H}_\mathbb{K})$ maximizes the weighted sum of spectral efficiencies for a given set of users $\mathbb{K}$. Therefore, the weighted sum of spectral efficiencies must be maximized by considering every possible choice of $\mathbb{K}$. That is, $\mathbf{p}_\theta^*(\mathbf{H}_\mathbb{K})$ must be computed for any of the $\sum_{m=1}^{M} \binom{K}{m}$ possible user sets in order to find $\mathbb{K}_\theta^*(\mathbf{H})$ in (6.4).

p0250   Note that MUD is exploited thanks to the exhaustive search among all possible user sets. Indeed, when the number of users in the system is high, it is very probable to find M non-interfering users so that the total throughput is approximately M times larger than that of an SISO system. Therefore, although the ZF beamformer is considered as a suboptimal strategy which does not achieve capacity, it can be shown that it is asymptotically optimal when the number of users in the system grows to infinity [4].

s0040

6.1.3 The average spectral efficiency region and the total average spectral efficiency

p0260   Combining the optimal power allocation with the optimal spatial multiplexing and scheduling, the boundary of the average spectral efficiency region can be obtained by means of the vector of relative priorities $\boldsymbol{\theta} = \{\theta_1, \ldots, \theta_K\}$. Summarizing Chapter 3, the average spectral efficiency region is defined as

$$\Omega = \left\{ \sum_{k=1}^{K} \theta_k \bar{R}_k \mid \theta_k \leq 1, \sum_{k=1}^{K} \theta_k = 1, \bar{R}_k \in S, k \in \{1, \ldots, K\} \right\} \quad (6.13)$$

where

$$S = \bigcup_{\mathbf{P} \in \mathbb{P}} \left\{ \bar{\mathbf{R}} : \bar{R}_k = E\left\{ R_k(\mathbf{H}, \mathbf{p}(\mathbf{H})) \right\}, k \in \{1, \ldots, K\} \right\} \quad (6.14)$$

and, therefore, the points at the boundary of Ω are obtained by governing the priority vector $\boldsymbol{\theta}$ in (6.2).

An example of the two-user average throughput region and of the two-user average information-theory rate region is given in Figure 6.2. Note that, whereas for the information-theory rate region the optimal power allocation is given by (6.10), for the average throughput region, the optimal power allocation has been approximated using (6.11). As noted all along in this book, the gap between the two regions is due to the packet error rate considered in the throughput measure in front of the error-free assumption

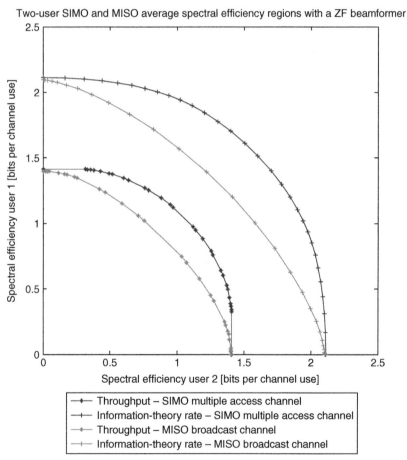

Figure 6.2 *Two-user MISO and SIMO average spectral efficiency regions with a ZF beamformer. Throughput is computed using the eight modulation and coding schemes with packet length equal to 1152 information bits, as in Table 2.3 in section 2.7. Power allocation for throughput in MISO broadcast channel is performed using* $\alpha_1 = 0.90$ *and* $\alpha_2 = 0.55$ *as in Figure 2.7. A Rayleigh fading channel with average SNR = 3 dB* $(P_{total} = 3\,dB, \sigma_2 = 0\,dB)$ *is considered. The number of antennas at the AP is two.*

in the information-theory rate measure. Furthermore, the average spectral efficiency region in the SIMO multiple access channel, as in Figure 5.2, is also plotted. We observe that, because of the power restrictions, the average spectral efficiency region is bigger in the SIMO multiple access channel than in the MISO broadcast channel. Particularly, in the former case, an average SNR of 3 dB corresponds to an *individual* power restriction of 3 dB and a noise power of 0 dB; in the latter case, an average SNR of 3 dB corresponds to a *total* power restriction of 3 dB and a noise power of 0 dB. Hence, these 3 dB of power are to be shared between the two users. As a consequence, when the priority vector is such that only one user can use resources ($\theta = \{1, 0\}$ or $\theta = \{0, 1\}$), the two regions coincide. However, as long as the priority vector is such that resources are to be distributed between the two users, the SIMO multiple access spectral efficiency region becomes bigger than that of the MISO broadcast channel at the expense of using twice the power.

p0280   The average spectral efficiency region is studied by maximizing the weighted sum of spectral efficiencies. Then, the system can work at any point on the boundary of the average spectral efficiency region by modifying the vector of relative priorities θ. Moreover, each user experiences an individual average spectral efficiency that depends on that vector of relative priorities. Throughout this book, the sum of all individual average spectral efficiencies is called the total average spectral efficiency and, as shown in previous chapters, the total average spectral efficiency is maximized when all users have the same priority.

p0290   In Figure 6.3, an example of the total average throughput when users have equal priority is presented. In this case, rather than presenting the differences between information-theory rate and throughput, which were already studied in previous chapters, the aim of the figure is to evaluate the advantages of the optimal power allocation with respect to the uniform power allocation. Therefore, the total average throughput with uniform power allocation is also plotted. Results show that differences between the optimal power allocation and the uniform power allocation strategy are very low. Hence, due to its simplicity, uniform power allocation is commonly chosen in practice.

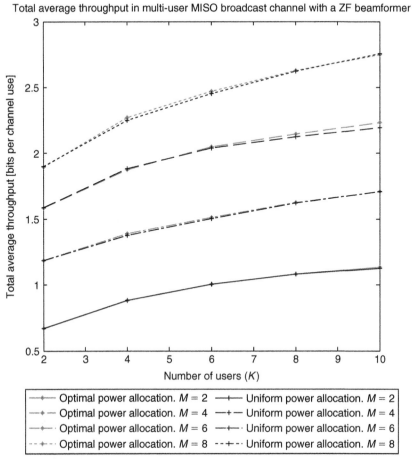

Total average throughput in multi-user MISO broadcast channel with a ZF beamformer

Legend:
- ─────── Optimal power allocation. *M* = 2 ──+── Uniform power allocation. *M* = 2
- ─ ─ ─ Optimal power allocation. *M* = 4 ──+── Uniform power allocation. *M* = 4
- ─·─·─ Optimal power allocation. *M* = 6 ──+─· Uniform power allocation. *M* = 6
- ─·─·─ Optimal power allocation. *M* = 8 ─·─+·─ Uniform power allocation. *M* = 8

Figure 6.3 *Total average throughput in multi-user MISO broadcast channel with ZF beamformer. Throughput is computed using the eight modulation and coding schemes with packet length equal to 1152 information bits, as in Table 2.3 in section 2.7. Rayleigh fading channel with average SNR = −3 dB.*

f0030

Nevertheless, even when considering uniform power allocation, the exhaustive search among all the $\sum_{m=1}^{M} \binom{K}{m}$ possible user sets is still to be performed. This is a computationally complex and very costly solution. Furthermore, full CSI information of all users in the system is needed which leads to a high amount of feedback data when the number of users is high. For these reasons, suboptimal but low complex spatial multiplexing and scheduling strategies will be considered in the next section.

s0050

6.2   Low complexity spatial multiplexing and scheduling policy

p0310

The optimal resource allocation solution obtained in the previous section presents two main problems that make its practical implementation difficult: first, the complexity of performing an exhaustive search over all possible sets of users; and second, the high amount of feedback information necessary to acquire the full CSI of all users.

p0320

One approach is to tackle these two drawbacks separately. On the one hand, low complexity suboptimal user selection algorithms have been studied [4, 7 and 8]. However, all these algorithms assume that the full CSI of all users is available at the resource allocation unit. On the other hand, the problem of limited feedback has been also studied [9, 10]. A common approach is to consider a constrained feedback channel where each user quantizes its channel vector to one of $N = 2^B$ quantization vectors and feeds back the corresponding index. This feedback is useful to obtain spatial direction of the channel and not the channel magnitude.

p0330

Another approach is the orthogonal random beamforming (or opportunistic beamforming) strategy [5]. In this scheme the access point selects a number of random beams and users feed back the highest SNR value that they receive on these beams. Then, the base station assigns each beam to the user with the highest SNR on that beam. The comparison between opportunistic beamforming and ZF beamforming is rather difficult because of the compromise between feedback and performance.

p0340

Following the same philosophy as in Chapter 5, in this chapter we will consider the less complex spatial multiplexing and scheduling algorithms with a ZF beamformer. With these low complex algorithms, the selection of users that is typically based on an exhaustive search is substituted by a random or round robin user set selection, thus reducing the scheduling complexity and the amount of feedback information.

p0350

Consider the MISO broadcast channel presented in Figure 6.1. Prior to the transmission of data, a set of users is selected as potential receivers. Then, the selected users feed back their CSI and a ZF beamformer is performed at the AP. In order to reduce complexity of the spatial multiplexing and scheduling policy, we assume that the decision on what is the set of receiving users is taken without the complete and instantaneous CSI, i.e. without the channel matrix **H**. However, once the set $\mathbb{K}$ of users is chosen, the channel matrix $\mathbf{H}_{\mathbb{K}}$ is estimated to perform a ZF beamforming. Note that a quantized feedback channel could be further considered in this second stage in order to reduce feedback information. However, the process of selecting users without the complete CSI implies already a huge amount of feedback reduction, especially when the number of users is high.

With this process, the resulting SNR for each user k in the set $\mathbb{K}$ is given by (6.5). Further, given the set $\mathbb{K}$, the optimal power allocation (6.3) could be considered. However, for the sake of low complexity, we will assume the uniform power allocation given by (6.12).

Similarly as in Chapter 5, we define a random policy characterized by a vector $\mathbf{t} = \{t_{\mathbb{K}} : \mathbb{K} \subseteq P\{1, \ldots, K\}\}$ with $t_{\mathbb{K}} \in [0,1]$ and $\sum_{\mathbb{K} \subseteq P\{1,\ldots,K\}} t_{\mathbb{K}} \leq 1$ where $t_{\mathbb{K}}$ is the probability that terminals in the set $\mathbb{K}$ are selected at a given time. Due to the fact that the set of active users $\mathbb{K}$ does not depend on **H** but on the policy **t**, the individual average throughput can be expressed as

$$\bar{R}_k = \sum_{\substack{\mathbb{K} \subseteq P\{1,\ldots,K\} \\ \text{s.t. } k \in \mathbb{K}}} t_{\mathbb{K}} E_{\mathbf{H}} \left\{ R^{(m^*)} PSR^{(m^*)}(\gamma_k^{ZF}) \right\} \qquad (6.15)$$

and the total average throughput is the sum of all the individual average throughputs given by $\bar{R} = \sum_k \bar{R}_k$.

Although the design of the random policy **t** accepts many different criteria depending on the designer, a simple and reasonable criterion would be to always spatially multiplex and schedule users in sets of equal size $\mathbf{K} = |\mathbb{K}|$. That is, to set the condition $\sum_{\mathbb{K} \subseteq P\{1,\ldots,K\} \ s.t. \ |\mathbb{K}|=\mathbf{K}} t_{\mathbb{K}} = 1$. In this case, the optimal set size that maximizes the total average throughput with no CSI can be obtained. Further, the amount of feedback information is always the same because the number of users that feedback CSI is always the same.

Then, depending on the limitations of the feedback channel, a maximum scheduling set size could be set.

As it was already justified in Chapter 5, for any random policy **t** such that $\sum_{\substack{\mathbb{K} \subseteq P\{1,\dots,K\} \\ s.t.\ |\mathbb{K}|=\mathbf{K}}} t_{\mathbb{K}} = 1$, the total average throughput is

$$\bar{R}_{\mathbf{K}} = \sum_{\substack{\mathbb{K} \subseteq P\{1,\dots,K\} \\ s.t.\ |\mathbb{K}|=\mathbf{K}}} t_{\mathbb{K}} \sum_{k \in \mathbb{K}} E_{\mathbf{H}} \left\{ R^{(m^*)} PSR^{(m^*)}(\gamma_k^{ZF}) \right\} \quad (6.16)$$

$$= \mathbf{K}\, E_{\mathbf{H}} \left\{ R^{(m^*)} PSR^{(m^*)}(\gamma_k^{ZF}) \right\}$$

because the expectation $E_{\mathbf{H}} \left\{ R^{(m^*)} PSR^{(m^*)}(\gamma_k^{ZF}) \right\}$ only depends on the set cardinality **K** as long as all users are homogeneous. The subindex **K** in (6.16) is used to indicate that the total average throughput obtained when users are spatially multiplexed and scheduled in groups of **K**, with independence on what is the set $\mathbb{K}$ that is chosen every time.

One objective of the designer would be to obtain the optimal value **K*** such that it maximizes the total average throughput given by (6.16). That is,

$$\mathbf{K}^* = \arg \max_{\mathbf{K}} \bar{R}_{\mathbf{K}} \quad (6.17)$$

Note that the optimization problem (6.17) refers to the broadcast channel but is the same as (5.12) which refers to multiple access channel. However, since power restrictions are different, the optimal solution is also different.

6.2.1   A closed form approximation for the average throughput in Rayleigh fading channels

A closed form approximation of $\bar{R}_{\mathbf{K}}$ would be of great help in order to solve the optimization problem (6.17). Note that with uniform power allocation and total power restrictions, the SNR is

$$\gamma_k^{ZF} = \frac{P_{total}}{\left\| \mathbf{v}_k \right\|^2 \sigma^2 \mathbf{K}} \quad (6.18)$$

where $\mathbf{K} = |\mathbb{K}|$ and the total power is equally distributed among all users in $\mathbb{K}$. Further, assuming link adaptation, we can define the parameter $\beta_{th}^{(m)}$ as the SNR enhancement necessary to achieve the minimum SNR $\gamma_{th}^{(m)}$ required for using the modulation and coding scheme m. Mathematically, this is expressed as $\gamma_{th}^{(m)} = \beta_{th}^{(m)} P_{total}/\sigma^2$ (a review of link adaptation was presented in Chapter 2).

Therefore, equally as in section 5.4.1, the total average throughput can be computed as

$$\bar{R}_{\mathbf{K}} = \mathbf{K} \sum_{m=1}^{Mod} R^{(m)} \int_{\mathbf{K}\beta_{th}^{(m)}}^{\mathbf{K}\beta_{th}^{(m+1)}} PSR^{(m)}(\beta) f_{\mathbf{K}}(\beta) d\beta \tag{6.19}$$

where Mod is the total number of modulation and coding schemes and the random variable $\beta = 1/[(\mathbf{H}_{\mathbb{K}}^H \mathbf{H}_{\mathbb{K}})^{-1}]_{kk}$ is a weighted chi-square distributed variable with $r_{\mathbf{K}} = 2(M - \mathbf{K} + 1)$ degrees of freedom and p.d.f. equal to $f_{\mathbf{K}}(\beta) = \beta^{K-\mathbb{K}} e^{-\beta}/(M - \mathbf{K})!$. The step function approximation of the PSR can be expressed in terms of α as

$$PSR^{(m)}(\beta) \simeq \begin{cases} 0 & \text{for } \beta \leq \mathbf{K}\beta_{th}^{(m)} \\ 1 & \text{for } \beta \geq \mathbf{K}\beta_{th}^{(m)} \end{cases} \tag{6.20}$$

Then, combining (6.19) and (6.20), it is easy to obtain

$$\begin{aligned} \bar{R}_{\mathbf{K}} &= \frac{\mathbf{K}}{(M - \mathbf{K})!} \sum_{m=1}^{Mod} \left(R^{(m)} - R^{(m-1)} \right) \Gamma\left(\frac{r_{\mathbf{K}}}{2}, \mathbf{K}\beta_{th}^{(m)} \right) \\ &= \mathbf{K} \sum_{m=1}^{Mod} \left(R^{(m)} - R^{(m-1)} \right) \sum_{i=0}^{M-\mathbf{K}} \frac{(\mathbf{K}\beta_{th}^{(m)})^i e^{-\mathbf{K}\beta_{th}^{(m)}}}{i!} \end{aligned} \tag{6.21}$$

where the gamma function $\Gamma(\beta, x)$ is $\int_x^\infty e^{-t} t^{\beta-1} dt$ and by definition, $R^{(m)} > R^{(m-1)}$ with $R^{(0)} = 0$.

Expression (6.21) is in the same form as (5.15). Therefore, differences between the theoretical approximation and simulations are very similar to those presented in Figure 5.6. Further, the general behavior with respect to the optimal resource allocation strategy will be also very similar to that presented in Figure 5.9. The main difference between (6.21) and (5.15) is that the factor **K** appears inside the gamma function and, hence, (6.21) accounts for the total power constraint. Some examples comparing the performance of low complexity spatial multiplexing and scheduling in MISO broadcast channels and SIMO multiple access channels are given in Figure 6.4 and Figure 6.5. Note that if the total power constraint in the MISO broadcast channel case was **K** (the number of scheduled users) times that of the individual power in the SIMO

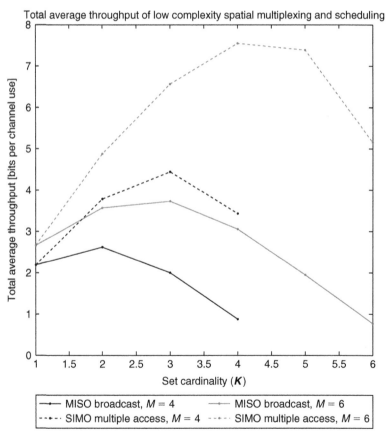

Figure 6.4 *Total average throughput of low complexity spatial multiplexing and scheduling in MISO broadcast and SIMO multiple access channels for different scheduling set cardinalities. Throughput is computed using the eight modulation and coding schemes with packet length equal to 1152 information bits, as in Table 2.3 in section 2.7. A Rayleigh fading channel with average SNR = 3 dB is considered.*

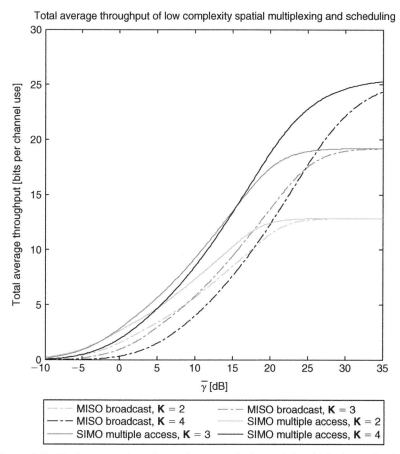

Total average throughput of low complexity spatial multiplexing and scheduling

⋯⋯ MISO broadcast, **K** = 2	⎯ ⋅ ⎯ MISO broadcast, **K** = 3
⎯ ⋅ ⎯ MISO broadcast, **K** = 4	⎯⎯ SIMO multiple access, **K** = 2
⎯⎯ SIMO multiple access, **K** = 3	⎯⎯ SIMO multiple access, **K** = 4

Figure 6.5 *Total average throughput of low complexity spatial multiplexing and scheduling in MISO broadcast and SIMO multiple access channels for different values of average SNR. A ZF beamformer is performed with M = 4 antennas. Throughput is computed using the eight modulation and coding schemes with packet length equal to 1152 information bits, as in Table 2.3 in section 2.7. A Rayleigh fading channel is considered.*

f0050

multiple access channel, the total average throughput would be the same in both cases.

6.3 Summary

The optimal resource allocation strategy that maximizes the weighted sum of spectral efficiencies in the multi-user MISO broadcast channel has been presented in this chapter. The maximization of the weighted sum of spectral efficiencies is interesting because the designer can decide at which point in the average spectral efficiency region it will work.

p0460 Further, we show that the optimal resource allocation can be divided into two optimization subproblems. One optimization problem is to find the optimal power allocation given a particular set of users. The optimal solution is presented as a water-filling strategy. Nevertheless, it is shown that the performance achieved by a uniform power allocation strategy is not far from the water-filling one. Consequently, the other optimization problem, which is to find the optimal set of users to whom power is to be allocated, is crucial. In that case the optimal solution is an exhaustive search strategy.

p0470 Since the exhaustive search strategy is a rather complex solution, low complexity spatial multiplexing and scheduling solutions are also presented. Following a similar methodology as in Chapter 5, a closed form solution of the total average throughput in a Rayleigh fading channel is obtained when an RR, or a random, spatial multiplexing and scheduling policy is implemented.

s0080
6.4 Annex: Water-filling algorithm with a ZF beamformer

p0480 A very simple algorithm exists when the water-filling solution is to be applied with a ZF beamformer. The main reason is that with a ZF beamformer the optimal power solution, (6.10) or (6.11), does not depend on the power allocated to the other users, as is the case if a MMSE beamformer is used [8]. Therefore, iterative algorithms similar to that in [8] can be further simplified.

p0490 Given the solution $p_k^* = a_k \left(\mu - b_k \right)^+$ with total power restrictions $\sum_{k \in \mathbb{K}} p_k^* = P_{total}$, an efficient algorithm to find the exact value for μ is described here. The objective is to find the water level μ for which the power constraints are satisfied with equality. To this end, we divide the problem into two steps. First, the users with some power allocated to them are found. Second, the exact water level and power levels are obtained.

p0500 To find the set of users in $\mathbb{K}$ with some allocated power, users are structured in increasing order of their parameters b_k so that the ordering vector $\boldsymbol{\pi} = (\pi_1, \ldots, \pi_{|\mathbb{K}|})$ is constructed with $b_{\pi_i} \leq b_{\pi_{i+1}}$. Then, the water level increased step by step so that starting from $\mu = b_{\pi_1}$, at the ith step μ is set to $\mu = b_{\pi_i}$. If at a given step the power sum of all users is less than the total power constraint, the next water level is used. This is repeated until a step is found such that more power than P_{total} is allocated. At this step, namely i, we can ensure that the final water level will be $\mu \in [b_{\pi_{i-1}}, b_{\pi_i})$ and the set of users $\mathbb{K}' = \{\pi_1, \ldots, \pi_{i-1}\}$ is such that $p_k^* > 0 \ \forall \ k \in \mathbb{K}'$. Note that $\mathbb{K}' \subseteq \mathbb{K}$. Then, the optimal water level is finely adjusted by means of $\sum_{k \in \mathbb{K}'} a_k (\mu - b_k) = P_{total}$ or equivalently,

$$\mu = P_{total} + \sum_{k \in \mathbb{K}'} a_k b_k / \sum_{k \in \mathbb{K}'} a_k.$$

All these steps are implemented in the following algorithm:

1. For $k \in \mathbb{K}$ compute $\|\mathbf{v}_k\|^2 = \left[\left(\mathbf{H}_{\mathbb{K}}^H \mathbf{H}_{\mathbb{K}} \right)^{-1} \right]_{kk}$ and set $a_k = \alpha_1 \theta_k$

   $\log_2 e$ and $b_k = \dfrac{\|\mathbf{v}_k\|^2 \sigma^2}{\alpha_1 \alpha_2 \theta_k \log_2 e}$

2. Order users in a vector $\boldsymbol{\pi} = (\pi_1, \ldots, \pi_{|\mathbb{K}|})$ such that $b_{\pi_i} \leq b_{\pi_{i+1}}$
3. Set $i = 1$
4. Set $\mu = b_{\pi_i}$, $i = i + 1$ and $p_k^* = a_k (\mu - b_k)^+$ for $k \in \mathbb{K}$
5. While $i \leq |\mathbb{K}|$ and $\sum_{k \in \mathbb{K}} p_k^* \leq P_{total}$ repeat step 4
6. Set $\mathbb{K}' = \{\pi_1, \ldots, \pi_{i-1}\}$
7. Set $\mu = P_{total} + \sum_{k \in \mathbb{K}'} a_k b_k / \sum_{k \in \mathbb{K}'} a_k$ and $p_k^* = a_k (\mu - b_k)^+$ for $k \in \mathbb{K}$

References

[1] H. Weingarten, Y. Steinberg and S. Shamai, The capacity region of the Gaussian multiple-input multiple-output broadcast channel, *IEEE Trans. Information Theory*, Vol. 52, pp. 3936–3964, 2006.

[2] M. Costa, Writing on dirty paper, *IEEE Trans. Information Theory*, Issue No. 3, pp. 439–441, May 1983.

[3] N. Jindal and A. Goldsmith, Dirty paper coding vs. TDMA for MIMO broadcast channels, *Proc. IEEE Int. Conference on Communications (ICC)*, 2004.

[4]   T. Yoo and A. Goldsmith, On the optimality of multiantenna broadcast scheduling using zero-forcing beamforming, *IEEE Journal on Selected Areas in Communications*, Vol. 24, Issue No. 3, pp. 528–541, March 2006.

[5]   M. Sharif and B. Hassibi, On the capacity of MIMO broadcast channels with partial side information, *IEEE Trans. Information Theory*, Vol. 51, Issue No. 2, pp. 506–522, Feb. 2005.

[6]   S. Boyd and L. Vandenberghe, *Convex Optimization*, Cambridge University Press, 2004.

[7]   C. Swannack, E. Uysal-Biyikoglu and G.W. Wornell, Low complexity multiuser scheduling for maximizing throughput in the MIMO broadcast channel, *Proc. Allerton Conference on Communications, Control and Computing (ACCC), 2004.*

[8]   M. Kobayashi and G. Caire, An iterative water-filling algorithm for maximum weighted sum-rate of Gaussian MIMO-BC, *IEEE Journal on Selected Areas in Communications*, Vol. 24, Issue No. 8, Aug. 2006.

[9]   N. Jindal, MIMO broadcast channels with finite rate feedback, *Proc. IEEE Globecom*, 2005.

[10] P. Ding, D.J. Love and M.D. Zolotowski, Multiple antenna broadcast channel with shape feedback and limited feedback, *IEEE Trans. Signal Processing*, Vol. 55, pp. 3417–3428, July 2007.

7

Different views of delay in resource allocation for wireless systems*

*This chapter is co-authored by Dr. Nizar Zorba.

Compared with a wireline backhaul, the wireless access link is actually the bottleneck in multi-user mobile communications, where several users communicate with a single AP. In fact, wireless channels are characterized by a harsh scenario for communications as the channel suffers from multiple undesired effects such as deep fades, multipath, etc., which make the information that reaches the receiver subject to different conditions. Therefore, huge efforts are being made to improve the wireless link capabilities by enhancing the wireless system performance and making it more suitable for commercial applications. Moreover, when considerating commercial applications in the wireless link, some QoS requirements must be guaranteed to the users of the system. In previous chapters, throughput was considered as the practical QoS metric to evaluate the system performance. In this chapter, the QoS metric under study will be the delay.

In packet-based communications (those considered throughout this book), end-to-end delay is the total time a packet takes from its source until it reaches its destination, where this total delay is the summation over a variety of delay sources in the system. A widely known source of delay is the queueing delay, characterizing the time invested in the queue of a transmitter or scheduler. Another source of delay strongly related to multi-user communications is the access delay that accounts for the time a user must wait until it gets access to the channel. In general, the more often a user gets access to the channel, the faster the packet delivery and, hence, the lower the packet delay. Clearly, this source of delay is largely affected by the resource allocation policy implemented at the AP where the resource allocation unit dynamically schedules transmissions.

p0030   Other sources of delay stem from the propagation delay or the packet processing delay at the transmitter and/or receiver side, but they are negligible when compared to queueing and access delays, so that they are not considered in this chapter.

p0040   Delay is actually one of the main performance indicators in the system, as some applications cannot wait a long time to receive their packets and there usually exists a delay threshold, beyond which packets are considered useless. Several kinds of thresholds are available on the basis of the system and/or application specification. These define the several types of delay measures that are present in the system, each one targeted to characterize a specific system behavior. The three main delay metrics are the instantaneous delay, the average delay and the instantaneous delay variation.

p0050   This chapter begins with the different metrics used to evaluate delay. Then, a discussion of the several sources of delay in the system is presented, where the performance of several scheduling policies is discussed in detail. Equal importance is given to both the queueing and access delay sources due to their impact on the system performance. The benefit from the availability of multiple antennas is also exposed in this chapter which underlines the schematics found in previous chapters.

s0010   ## 7.1   The delay metrics

p0060   The sources of delay have a wide impact on the system performance but also the delay metrics that are employed to evaluate the system performance show great importance in the system. Indeed, the delay can be measured using different metrics depending on the application under consideration, where an application can be more sensitive to a given delay measure while other applications might be more stringent on another delay metric. Although there might exist many different criteria to evaluate the delay in a communications system, in this section we present the most representative metrics: the instantaneous delay, the instantaneous delay variation, the average delay and the worst-case delay with each one defining a certain system behavior.

7.1.1 The instantaneous delay

The system applications usually have maximum deadlines for the reception of their packets, for example when some user is running a voice application, voice packets have to arrive within a maximum time threshold to be correctly decoded. This is required because in voice applications it is preferable to receive a discontinuous voice flow with some losses but low delay than a continuous voice flow without losses but with a high delay.

Hence, several applications require a maximum instantaneous delay D_{max} for their proper operation, standing as a hard restriction on the system performance, where the joint effect of all the delay sources has to be controlled and optimized to guarantee the satisfaction of the maximum delay deadline. In multi-user communications, a tool that can be used to perform this optimization is the connection admission control (CAC) unit to control the total number of users K in the system [1] such that at every time instant t

$$\max \ K$$
$$\text{s.t. } D_k(t) \leq D_{max} \ \forall k \in \{1, ..., K\} \qquad (7.1)$$

with $D_k(t)$ as the instantaneous packet delay for each user in the system. Note that, as the chapter focuses on packet-oriented transmissions, the time index t is discrete. Maximum instantaneous delay is of great importance in voice applications, but also in any other interactive application like video calling, online gaming, etc.

7.1.2 Instantaneous delay variation

Another important metric to account for the delay performance is the instantaneous delay variation defined as the time difference between the reception of two consecutive packets [2]. This metric is of special interest to some applications where a regular delivery of packets is desired. More specifically, in voice applications, packets must arrive with the same instantaneous packet delay to make the communication seem smooth to the final user. In certain commercial systems, the playout buffers can be employed to partially compensate for the delay variation [3]. In this case, the maximum delay variation is of special interest in order to size the playout buffers.

p0100   As already noted in [2], the variation in packet delay is sometimes also known as jitter, but the word jitter has several meanings, for example it can represent the input signal variation with respect to a given clock (e.g. synchronized communications), but also it can refer to the signal variation with respect to some metric (e.g. variation with respect to the average delay). To avoid this confusion in this chapter, the word jitter is not used.

s0040   ### 7.1.3 *Average delay*

p0110   Other applications do not require instantaneous delay for their correct operation but the main target is to receive the largest amount of data over time. Therefore, the average delay [4] is what defines the system performance. An example is any file transfer application where the main restriction is not the maximum instantaneous delay but the average delay, so that the system needs to get all the data within the minimum time interval.

p0120   The average delay is actually the mean value of the instantaneous packet delays over a long time interval. Therefore, the mathematical expression for the average packet delay $\bar{D}_k$ of user k, measured in time slots, states that

$$\bar{D}_k = \lim_{T \to \infty} \frac{\sum_{t=0}^{T} D_k(t)}{T} \tag{7.2}$$

with T as the time interval where the instantaneous packet delay is averaged.

p0130   In comparison to the instantaneous delay, the average delay is the long-term delay metric obtained over a long time interval, while the instantaneous delay is denoted as the short-term delay as it is measured over a single channel realization.

p0140   To clarify the difference between average delay, instantaneous delay and instantaneous delay variation, Figure 7.1 shows two situations of different packets arriving at destination. In both

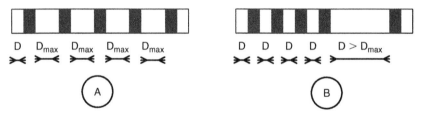

Figure 7.1 *Two situations for the packets arrival process.*

cases the average delay is the same, but for case A all the packets are received within a maximum instantaneous delay restriction, while for case B, the first four packets satisfy the maximum
instantaneous delay restriction, while the last (fifth) packet does
not. Furthermore, note that in case A the time between the arriving
packets is the same, showing a null variation in this case. In case
B, as the last packet is received with a large time difference with
respect to all the preceding packets, the system delay variation is
increased.

7.1.4 *Worst-case delay*

Another delay metric available in the literature [5] is the worst-
case delay. This metric is specific for multi-user communications
and is defined as the invested time by the resource allocation unit
to award the service to every one of the users in the system. This is
actually the hardest delay metric as all users must receive access to
the channel, regardless of their channel, queue characteristic and
traffic nature. This metric can be very interesting for operators that
require a fair resource distribution among their customers.

7.2 Sources of delay

This book deals with packet-oriented communications where
packets are originated from a source and travel through the system
until they reach their destination. Through this 'travel' process,
there are many different sources of delay along the path between
the source and the destination of a packet and, hence, the end-to-
end delay can be seen as a summation of many different delays. In
multi-user communications, the main two sources of delay are the
access delay and the queueing delay, as shown in Figure 7.2.

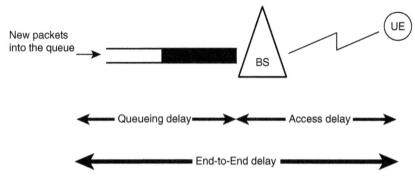

Figure 7.2   *The delay sources in the system.*

p0170   The impact of each of these sources is different and depends very much on the application and/or scenario conditions, so that in single user wireless systems, the largest source of delay is the queueing delay, while for multi-user wireless scenarios running delay-strict applications, the access delay also has a very strong impact. The study of the end-to-end delay in multi-user wireless networks is a very complex topic; however, to provide the reader with a general insight into the impact that the different delay sources have on the end-to-end delay, this chapter separates the access delay and the queueing delay, and individually studies their impact and performance.

p0180   Now, a small introduction about the two major sources of delay is presented to put the reader in the context of system delay and to realize their application scenario, while in the following sections (7.3) and (7.4), a detailed study for each source is individually performed, with special interest in the resource allocation process.

s0070   ## 7.2.1 The access delay

p0190   In the uplink, the access delay is defined as the time between two consecutive channel accesses of the same user. In fact, it is the time a user must wait from one channel access to the next one. Conversely, in the downlink, the access delay is defined as the time taken by the AP between two consecutive channel awards for the communication with the same user. Whether the chapter refers to the uplink or downlink of the communication system, the access delay can be seen as the time between two consecutive allocations of resources to the same user.

Clearly, the access delay has a strong impact in multi-user communication systems where all users cannot access the channel at the same time. The resource allocation unit is the one in charge to schedule the users' transmissions either in the uplink or in the downlink. Therefore, the access delay is also commonly known as scheduling delay.

The access delay is strongly related to the instantaneous delay and the instantaneous delay variation metrics as the resource allocation unit is concerned with awarding the channel access for each user/s on each time instant. Therefore, the access delay is typically measured in terms of instantaneous delay and delay variation.

7.2.2 The queueing delay

Another important delay source in the system relates to the delay that the packets suffer in the transmitter queues. The generation of the packets (data downloaded from the Internet, voice encoder, etc.) follows some statistical behavior at the information source that does not match the channel statistics. Therefore, some queues are required at the transmitter side to keep the packets before their transmission through the wireless channel. The time that a packet waits in the system queue until that packet enters the scheduling unit is defined as the queueing delay, and it mainly depends on the source statistics and the service time, which is defined as service time = access delay + propagation delay + packet processing delay + automatic repeat request (ARQ) delay. As the propagation and processing delays are neglected, together with a non-retransmission policy, so that no ARQ is possible, then service time and access delay converge to the same value.

7.3 Access delay on resource allocation

Throughout this book, the general problem of resource allocation in wireless communication systems has been tackled by considering throughput as a figure of merit. Then, the total average throughput in SISO systems has been studied by analyzing the performance of MUD scheduling. Similarly, in SIMO and MISO systems, the total average throughput maximization problem has been formulated as a simplified spatial multiplexing and scheduling problem where, every time instant, the optimal set of simultaneously transmitting users is to be scheduled. Moreover, as this optimal solution is shown to be computationally demanding, low complex spatial multiplexing and scheduling policies such as RR scheduling and random scheduling have been analyzed also. Therefore, several resource allocation techniques have been presented in terms of throughput.

p0240   But now, a discussion on the performance of each scheduling philosophy is presented in terms of the access delay, with each one of them showing some benefits/disadvantages depending on the system objectives and restrictions.

s0100   ## 7.3.1 *RR scheduling*

p0250   The RR policy is the easiest approach to deliver service to users, so that users are periodically selected through a predefined order [6]. Each one of the users knows in advance when it will have access to the channel and, therefore, the access delay is tightly obtained [5]. The average delay of this scheduling policy is the same as that of the random scheduling policy when all users have the same probability of being scheduled, but they provide different instantaneous access delay results. Furthermore, the RR scheduling philosophy is optimal in terms of instantaneous access delay. In particular, assume that the number of users in the system is denoted by K and that, every time slot, users are scheduled in groups of $\mathbf{K}$ following a round robin fashion. Then, the instantaneous access delay of any user k in the system, and measured in time slots, is computed as

$$AD_k^{RR}(t) = \frac{K}{\mathbf{K}} \tag{7.3}$$

p0260   Note that because the RR strategy does not introduce any randomization on the service provision, the average access delay of any user k in the system is exactly the same. That is,

$$\overline{AD_k}^{RR} = \frac{K}{\mathbf{K}} \tag{7.4}$$

p0270   Round robin is the simplest technique for scheduling users, and it is currently implemented in TDMA systems like GSM. This is mainly due to its low complexity design and guaranteed users' access to the channel. Moreover, it shows a null instantaneous delay variation, as the time between consecutive channel accesses is fixed forever.

0280   Clearly, the minimum access delay would be achieved by scheduling users in groups of $\mathbf{K} = K$ users. That is, simultaneously scheduling all users. However, the performance in terms of spectral efficiency might be very low as users might create too much interference among them. For instance, in multi-user SIMO and MISO systems with a ZF beamformer, there exists an optimal scheduling set size, given by (5.12) and (6.4), respectively, which maximizes the total average spectral efficiency.

0290   Alternatively, the designer could be interested in finding the optimal scheduling set size $\mathbf{K}^*$ such that the access delay is minimized while a minimum total average spectral efficiency $\bar{R}_{\min}$ is guaranteed. Therefore, the objective would be

$$\mathbf{K}^* = \max_{\substack{\mathbf{K} \\ s.t. \ \bar{R}_{\mathbf{K}} \geq \bar{R}_{\min}}} \mathbf{K} \tag{7.5}$$

0300   Summarizing the results in Chapters 5 and 6, when the scheduling set size is $\mathbf{K}$, the total average throughput in Rayleigh fading channels is very well approximated by the closed form expressions

$$\bar{R}_{\mathbf{K}}^{SIMO} = \mathbf{K} \sum_{m=1}^{Mod} (R^{(m)} - R^{(m-1)}) \sum_{i=0}^{M-\mathbf{K}} \frac{(\alpha_{th}^{(m)})^i e^{-\alpha_{th}^{(m)}}}{i!} \tag{7.6}$$

$$\bar{R}_{\mathbf{K}}^{MISO} = \mathbf{K} \sum_{m=1}^{Mod} (R^{(m)} - R^{(m-1)}) \sum_{i=0}^{M-\mathbf{K}} \frac{(\mathbf{K}\alpha_{th}^{(m)})^i e^{-\mathbf{K}\alpha_{th}^{(m)}}}{i!} \tag{7.7}$$

where *Mod* is the total number of modulation and coding schemes, $R^{(m)}$ is the effective transmission rate in modulation and coding scheme m, the number of antennas at the AP is given by M and $\alpha_{th}^{(m)} = \gamma_{th}^{(m)}/\bar{\gamma}$ is the SNR enhancement necessary to achieve the minimum SNR required for the modulation and coding scheme m, when the average SNR is given by $\bar{\gamma} = P/\sigma^2$ in the SIMO case and by $\bar{\gamma} = P_{total}/\sigma^2$ in the MISO case.

p0310   In consequence, there is a trade-off between access delay and throughput because from (7.3) the access delay is reduced by increasing **K** but from (7.6) we observe the fact that by increasing **K** does not necessarily mean that throughput is increased. An example of such a trade-off is given in Figure 7.3 where the total average throughput given by (7.6) is plotted as a function of the scheduling set size cardinality **K**. This figure corresponds to one of the curves in Figure 5.7 from Chapter 5. Note that due to restrictions on the ZF beamformer, the maximum scheduling set size is **K** = *M* and, therefore, the minimum access delay is *K/M*. A similar example could be obtained for the MISO case.

p0320   The RR strategy is a low complexity resource allocation strategy that allocates resources without considering CSI in the users' scheduling. On the one hand, due to its deterministic behavior, RR

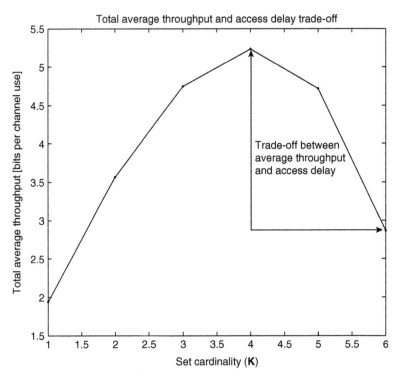

f0030   **Figure 7.3** *Total average throughput and access delay trade-off with M = 6 antennas. Throughput is computed using the eight modulation and coding schemes with packet length equal to 1152 information bits, as in Table 2.3 in section 2.7. A Rayleigh fading channel with average SNR = 0 dB is considered.*

scheduling minimizes the access delay, not only in terms of instantaneous packet delay but also in terms of the instantaneous packet delay variation. On the other hand, this strategy is a suboptimal strategy in terms of total average throughput because MUD cannot be exploited without CSI. But due to the increased demand for high spectral efficiency in current multi-user systems, then other sophisticated scheduling algorithms that can account for the scenario conditions replace the RR. Moreover, as the RR scheduler does not consider the users' CSI characteristics, then the access can be awarded to users with deficient channel conditions, making the reception incorrect and causing a packet drop. Therefore, the CSI consideration is required for a better system performance. Next, the access delay performance of spatial multiplexing and scheduling with CSI is studied.

7.3.2   Spatial multiplexing and scheduling with CSI

The philosophy of this resource allocation strategy is totally opposite to the RR one, as the users with the best channel characteristics are selected at the expense of randomizing users' channel access times. The AP detects an opportunity when a set of users shows good channel conditions and decides to benefit from this fact by selecting those users to simultaneously transmit data. In multi-user SISO systems this strategy is also known as the MUD opportunistic scheduling [7, 8]. A complete discussion of this technique is available in Chapter 4 and since multiple antennas introduce an additional spatial dimension, the MUD strategy is extended to a spatial multiplexing and scheduling strategy [9, 10] in Chapters 5 and 6.

Following the description in Chapters 5 and 6, the spatial multiplexing and scheduling strategy that maximizes the sum of spectral efficiencies aims at scheduling the optimal set of users $\mathbb{K}^*(\mathbf{H})$ such that

$$\mathbb{K}^*(\mathbf{H}) = \arg \max_{\mathbb{K}} \left(\sum_{k \in \mathbb{K}} R_k(\mathbf{H}_{\mathbb{K}}, \mathbf{p}(\mathbf{H})) \right) \tag{7.8}$$

where $\mathbf{H}$ is the matrix defining the channel conditions, $\mathbf{H}_K$ denotes the channel matrix when K is the set of scheduled users and

$R_k(\mathbf{H}_\mathbb{K}, \mathbf{p}(\mathbf{H}))$ is the spectral efficiency of user k. Recall that in the SIMO system, such spatial multiplexing and scheduling policy is equivalent to a power allocation policy [11] when the power vector $\mathbf{p}(\mathbf{H}) = \{p_1, ..., p_K\}$ is set to $p_k = P$ for all $k \in \mathrm{K}$, and $p_{k'} = 0$ for all $k' \notin \mathrm{K}$. In the MISO case, the power allocation is given by the solution of (6.3).

p0350　　In general, a high number of users in the system are beneficial to spatial multiplexing and scheduling with CSI as the scheduler can choose among a larger pool of users [9]. Unfortunately, spectral efficiency improvements come at the expense of access delay performance because the best users are always selected, so that no user can have any guarantee for instantaneous service. Indeed, as the users' selection procedure is based on the channel characteristics, and as the channel is random and fluctuates over time, users are not periodically serviced and the instantaneous access delay variation is increased.

p0360　　Therefore, the instantaneous access delay of spatial multiplexing and scheduling with CSI is not tightly formulated, which is a required measure for several applications, being a drawback for the commercial implementation of this scheduler, even of its rate optimality [12]. Moreover, as best users are always selected, users with very bad channel conditions are never selected and some unfairness among users appears. Clearly, the fairer scheduling policy is the RR scheduling as all users are awarded the same access delay. To obtain some of the benefits of spatial multiplexing and scheduling with CSI, while at the same time having some control over the access delay as in the RR scheduling, two intermediate schedulers are now presented: the proportional fair scheduling and the frame division scheduling.

s0120　　### 7.3.3　Proportional fair scheduling

p0370　　To guarantee the access to the users with bad channel conditions, a weighted version [8] of the spatial multiplexing and scheduling with CSI is proposed, where the users showing best channel conditions with respect to their own channel statistics are selected. This scheduler, which is commonly identified as the proportional fair scheduler [8], guarantees that all users in the system are regularly scheduled, while at the same time some benefit from the channel characteristics is extracted. Therefore, the access delay unfairness of spatial multiplexing and scheduling with CSI is partially compensated. This scheme was initially proposed for multi-user SISO systems motivating its commercialization in UMTS-HSDPA and Qualcomm-HDR standards for cellular communications.

Within this approach, the scheduler accounts for the average channel indicator (e.g. average spectral efficiency) for each user and normalizes the instantaneous measure by the average value, to make the user selection process as

$$\mathbb{K}^*(\mathbf{H}) = \arg \max_{\mathbb{K}} \left(\sum_{k \in \mathbb{K}} \frac{R_k(\mathbf{H}_\mathbb{K}, \mathbf{p}(\mathbf{H}))}{\bar{R}_k} \right) \qquad (7.9)$$

with $\bar{R}_k$ as the average spectral efficiency for user k. But as the calculation of the average channel indicator is causal, instead of $\bar{R}_k$, an approximate average indicator $\bar{R}_k(t)$ is obtained through a moving average indicator that is formulated as

$$\bar{R}_k(t+1) =$$
$$\begin{cases} (1 - 1/\rho)\bar{R}_k(t) + (1/\rho)R_k(\mathbf{H}_{\mathbb{K}^*}, \mathbf{p}(\mathbf{H})) & \textit{if } k \in \mathbb{K}^*(\mathbf{H}) \quad (7.10) \\ (1 - 1/\rho)\bar{R}_k(t) & \textit{if } k \notin \mathbb{K}^*(\mathbf{H}) \end{cases}$$

where $R_k(\mathbf{H}_{\mathbb{K}^*}, \mathbf{p}(\mathbf{H}))$ is the spectral efficiency of user k if user k is in the optimal set $\mathbb{K}^*(\mathbf{H})$ given by (7.9). Recall that the dependence of $R_k(\mathbf{H}_{\mathbb{K}^*}, \mathbf{p}(\mathbf{H}))$ in time is through the channel matrix $\mathbf{H}$. In fact, $\mathbf{H}$ is employed to model a time varying block fading channel. Hence, the discrete time index t refers to the time slot where a given channel realization $\mathbf{H}$ is assumed to be constant.

This scheduler actually performs an exponential low pass filtering over the spectral efficiency values for all the users in the system. This process makes the user that has been given the access in the time instant t to decrease its probability to be awarded the channel access in time instant $t + 1$. The amount of decrease depends on the window length, defined through the parameter ρ, which also controls the access delay in the system.

p0400   A very large ρ makes the proportional fair scheduling converge
to the spatial multiplexing and scheduling policy as the influence
of the previously awarded rates is decreased, thus the scheduler
has more freedom to select the best users at each time instant [8].
On the other hand, a value $\rho \rightarrow 1$ makes the gains obtained by
using CSI disappear and then all users have the same probability
of being selected; making the proportional fair strategy a random
scheduler. As previously mentioned, the difference between the
random scheduler and the RR one, for the $\rho \rightarrow 1$ case, is that the
users are not selected in a sequential manner, which is, in fact, a
major difference to the RR philosophy where each user is guaranteed periodic access to the channel. Obviously, on average, the
performance between proportional fair with $\rho \rightarrow 1$ and RR is
equivalent, but for instantaneous access delay consideration, the
application of the moving average indicator makes the proportional fair scheduler differ slightly to the RR philosophy. Through
the larger probability to access the channel, within a given time
interval for every user in the network, the instantaneous access
delay variation is also improved as each user is serviced more
periodically. Closed form expressions for the access delay with
respect to parameter ρ are not available in the literature, so that the
adjustment of this parameter is usually performed by the operator
in commercial systems.

p0410   Other proposals to carry out the weighting in (7.9) are presented in
the literature, where [13] shows a utility-based weighting process
while [14] provides a weighting scheme when users do not inter-
fere with one another.

p0420   The benefits in terms of instantaneous access delay and delay
variation come at the expense of a reduced average spectral efficiency, where the spectral efficiency for proportional fair scheduling depends greatly on the weighting value ρ when compared to
the one corresponding to the spatial multiplexing and scheduling
policy. The average access delay is also improved by the proportional fair scheduling but, as just mentioned, the parameter ρ does
not provide an exact control over the instantaneous access delay
and, therefore, the instantaneous access delay is not tightly guaranteed to the users. Next, an alternative scheduling policy is presented where the weighting is performed through a frame division,

so that it can be considered as a practical implementation of the proportional fair approach, but with a tight control on the access delay.

7.3.4 Frame division scheduling

To benefit from the spectral efficiency increase of spatial multiplexing and scheduling with CSI and at the same time to guarantee tight instantaneous access delay to users, the scheduler can carry out a frame division [15]. In the first part of the frame, users are scheduled through an RR manner, so that they are all guaranteed to access the channel, thus controlling their access delay. Within the second part of the frame, the users with best channel conditions are selected, boosting the global system spectral efficiency performance. The frame division (FD) scheduling offers the advantage of tight access control with respect to the proportional fair scheduling, as the percentage of RR in each service slot is exactly known, so that the access delay values are extrapolated from the RR philosophy.

Through the frame division, the achievable spectral efficiency in the system is increased in comparison with that of the RR scheme, obviously at the cost of larger access delay and access delay variation. This merging of the two scheduling philosophies is very attractive, providing a trade-off over the two system requirements: access delay and average spectral efficiency. The operating point in the trade-off is controlled through the design of the percentage of each division in the frame, so that, for example, 60% of the frame can be awarded to the access delay requirements, while the remaining 40% is allocated for spatial multiplexing and scheduling with CSI. This is graphically shown in Figure 7.4.

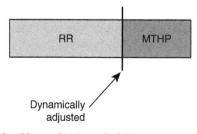

Figure 7.4 *An example of frame division scheduling.*

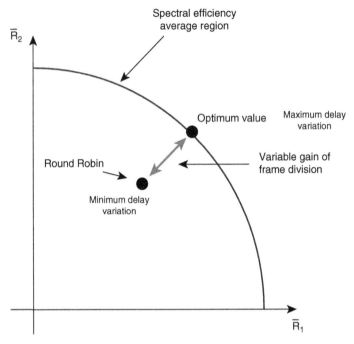

Figure 7.5   *Spectral efficiency region for the frame division scheduler. The optimum value is achieved with spatial multiplexing and scheduling with CSI.*

p0450   Figure 7.5 gives more insight into the spectral efficiency of the FD scheme, while a comparison with the spatial multiplexing and scheduling with CSI scheduler is presented, in terms of their average spectral efficiency. Depending on the frame percentage that is awarded for access delay, the FD gain in terms of the average spectral efficiency varies, where the two extremes are obviously defined by the RR point and the optimum value for the spatial multiplexing and scheduling with CSI scheduler. As with all the previous scheduling policies, the users' access delay is characterized, with each one of the scheduling techniques offering advantages and drawbacks for their implementation, so that the system architecture engineer can consider the scheme that best matches with the system requirements and restrictions.

s0140   ## 7.4   Queueing delay on resource allocation

p0460   For all applications that require finite capacity resources for their operation (actually almost all practical systems), the system also requires some queues at the transmitter side. The packets' arrival is usually unpredictable and the wireless channel characteristics are random; thus, there does not exist a match between the packets' arrival rate and the service rate, and therefore some mechanism to keep the packets before their transmission is required. These packets are kept in queues upon their arriving time, and the waiting time for each packet within these queues until that packet enters the scheduling unit is defined as the queueing delay.

470 Two main factors are necessary for the queues' existence: the unpredictable arrival process and the random channel characteristics. Notice that the first factor is common to computer, wireline, wireless, single user and multi-user systems, which explains the huge contributions [16, 17] on queueing delay, but for the applications within the wireless systems, the second factor of the random channel characteristics plays a major role, complicating the study of the queueing systems in wireless channels, and explaining the few contributions on queueing delay for the wireless systems. These handicaps are even more important in wireless multi-user systems, with each user showing different channel characteristics. As the transmitter can provide service to several users at the same time, with interference among the serviced users, the delay study is further complicated.

480 The relation between the arrival rate and the service rate is crucial for the system operation, as an average arrival rate much lower than the average service rate makes the system little used, so that the system is inefficient in its usage of the available resources. On the other hand, an average arrival rate higher than the average system rate makes the queues increase without bounds, driving the system towards instability. In the last section of this chapter, a discussion about the system stability is presented. Notice that the queueing delay is closely related to the average delay metric, as both the arrival and service rates are presented in terms of the average measures.

7.4.1 Queueing delay parameters

490 The queueing delay depends on several system specifications affecting the time each packet remains in the system queue and for each one of the users. These factors [16] are: the arrival process, the

service time distribution, the available number of links, the queue's maximum allowed length, the total number of users in the system and the service policy. Each of these concepts is now discussed.

7.4.1.1 The arrival process

The information packets are generated at a data source where the source statistics differ among the several applications, users and/or environments; each one defined by some distribution for the packets' arrival process. This characterization plays a major role in the queueing delay, as it defines the average packets' load that comes from the source and is stored at the system queue, thus denoting the input to the queue. Obviously, the more packets the source generates, the higher the queueing delay each packet suffers in the system, where the packets arrival process $a_k(t)$ for the kth user is defined as

$$a_k(t) = \sum_{p=1}^{P_k(t)} b_p \tag{7.11}$$

with $P_k(t)$ as the number of arriving packets for a given time instant while b_p defines the number of bits in each packet. Both $P_k(t)$ and b_p can be described through several distributions, thus defining a global distribution for the arrival process, which can follow a Poisson, Markov, deterministic, exponential or general distribution, among others. The value of $a_k(t)$ also describes the interarrival time between the packets, which mainly accounts for several scenarios:

1. A memory-less system with independent and identical packet distributions, where the distribution of $a_k(t)$ follows a Poisson function.

2. Future input values depending on the current situation with the arrival process being defined through an exponential distribution; this is modeled by a Markov chain.

3. A special Markov case when the future values can be only close (i.e. neighbor Markov states) to the current situation as denoted by a birth–death process.

7.4.1.2 Service time distribution

Service time distribution represents the amount of packets that can be serviced to the user per each time interval, defining the output of the system queue. Even the service time can be known for certain scheduling schemes, but the characterization of the service rate is very difficult for the wireless channels, due to the multipath, fading and the random channel fluctuations, therefore being a challenging aspect to quantify the queueing system delay. In all other scenarios (computer, wireline, etc.) the queueing delay has closed form expressions, but the already mentioned handicaps for communication through the wireless link make the queueing delay for the wireless channels, and more specifically for the multi-user wireless channels, more difficult to formulate. The distribution for the wireless channels is denoted with a general distribution as its characterization is not known, but, obviously, if the channel can absorb a larger number of packets, then the queueing delay decreases.

7.4.1.3 *The available number of links*

Another important factor to determine the queueing delay is the number of links that are available for the transmitter side, where any multiple access scheme is always beneficial, as several packets can travel in the wireless channel at the same time. Therefore, code division multiple access (CDMA), frequency division multiple access (FDMA) and space division multiple access (SDMA) can substantially increase the output from the queue, thus motivating their use in the wireless channel to compensate for the bad channel performance, which is why special attention has been given to the role of SDMA throughout this book.

7.4.1.4 *Queue's maximum allowed length*

This parameter also reflects the number of accepted users (K), as a high number of accepted users in the system translates into a large number of information sources to the system queue, thus increasing the queueing delay for all the users. The number of users actually defines the system stability, as the system cannot account for a very high number of users because the average arrival rate would be larger than the average service rate, thus the packets remain in the system making the system queue increase until infinity, causing system instability [18]. To satisfy the maximum queue restriction, an optimization of the number of accepted users is usually carried out by a CAC unit [1, 19], where a short allowed queue size would accept a lower number of information sources as compared to a system that enables a large queue.

s0200

p0570

7.4.1.5 *The total number of users in the system (Nt)*

The CAC can choose the selected users among the total number of users in the system, but the total number of available users can be lower than the optimum number of users in the system, thus giving more importance to the parameter Nt. Notice that a smaller number of users in the system always induce a lower queueing delay for all the system, but a large number of serviced customers (users) are always beneficial to the operators, as income is increased. Therefore, a trade-off between the two effects is generated, which motivates the requirement for the CAC unit in the system.

s0210

p0580

7.4.1.6 *The service policy*

In the first part of this chapter, the access delay was presented where the scheduling policy that is employed by the system plays a major role in defining the users' access delay. Now, in the section of queueing delay, another form of scheduling, but now among the packets present in the queue, also has an important impact over the queueing delay. As there exist several packets in the queue, some policy is required to select the packet that will be the first for transmission (i.e. the one that enters the access delay measurements). This policy is actually defined through the service policy implemented over the queue. Even it seems logical to think that the first packet in the queue is the one that has to enter the service, but several policies are proposed (and implemented) to account for different restrictions and system requirements [16, 17], for example: First Come First Serve (FCFS), Last Come First Serve (LCFS), Shortest Processing Time First (SPT) and Biggest In First Served (BIFS), among others. All these serving philosophies are not channel aware policies, so that the packet is selected from the queue without any regard of the user channel characteristics. In the last section of this chapter, when presenting the stability operating conditions, the optimal service policy is shown to be channel aware in selecting the packet from the queue at the same time as the user with best channel conditions [18].

p0590

Notice that in multi-user systems, each packet from the source to its destination goes through two selection procedures: the service policy at the system queue and the scheduling step to select the user who accesses the channel, where each procedure has its own targets and restrictions.

The packets from the different sources can be put all together in a joint queue for all the system, as previously discussed in this section, but it can be also put in several queues, for example a single queue for each application and/or for each accepted user in the system.

An important aspect regarding the two already discussed types of delay (access and queue delays) is their relation to the throughput concept that was discussed in a previous chapter. A system that shows a high throughput for its wireless channel, then a larger number of packets, can be transmitted and the system queueing delay decreases for that system. Therefore, the throughput is beneficial for the queueing delay. On the other hand, for some scheduling philosophies, the increase in the system throughput comes at the expense of the access delay. For example, in the spatial multiplexing and scheduling with CSI policy, increasing the number of available users improves the system throughput [9], but the access delay is then increased as each user has to wait for a longer time until it is again serviced when K is large [19]. Therefore, a tradeoff appears between throughput and access delay, making the throughput increase harmful for the access delay.

7.4.2 Queueing delay and stability consideration

Considering an individual queue for each user in the system with a maximum queue length of $Q_{max}(k)$, then if more packets are available in the system, the extra packets are dropped, for what is defined as the queue overflow. A system is defined as a stable system if $Q_k < Q_{max}(k)$, which is a necessary requirement to avoid an infinite queueing delay in the system. Clearly, by guaranteeing that a steady-state distribution on Q_k exists with the property $Pr(Q_k > Q_{max}(k)) \to 0$ as $Q_{max}(k) \to \infty$ then the system is stable and, hence, queues in the system do not increase until infinity.

With all the previous parameters, the system queue shows a dynamic behavior due to its inputs from the information sources and the outputs to the channel. The queue dynamics, with an individual queue for each user in the systems, is defined as

$$Q_k(t+1) = \left[Q_k(t) - T \sum_{k=1}^{K} R_k(\mathbf{H}_{\mathbb{K}}, \mathbf{p}(\mathbf{H})) \right]_+ + a_k(t) \qquad (7.12)$$

where T is the service time duration and $R_k(\mathbf{H}_K, \mathbf{p}(\mathbf{H}))$ is the spectral efficiency that the serviced user k obtains during a period of time T. To avoid a negative queue length, $[v]_+$ is used and stands as $\max(v, 0)$, where $Q_k(t)$ defines the current queue length for the kth user. Notice that for a given stationary power allocation policy, the queue length at $t + 1$ is independent of the past time instants. Hence, the system can be viewed as evolving according to a Markov chain on a K-dimensional uncountable infinite state space.

p0640   For the stability requirement, the system is asked to be always stable in the long term, so that the average arrival rate is lower than the average service rate. An important queue characterization is thus denoted in terms of the maximum average delay that the network can support, before getting into system instability and causing the queues to blow up.

p0650   The set of average input rates under which there exists a power allocation policy that makes the system stable is called the stability region. Interestingly, the stability region and the spectral efficiency region overlap [21] and the optimal power allocation policy that uses the queue lengths Q_k to define terminal priorities is stable as long as the average input rates are inside the average spectral efficiency region. This is of special interest in systems where the average input rate is not known.

p0660   Therefore, the stability optimal scheduling policy accounts for the queue condition besides the channel characteristics [18]. The weighted sum maximization presented in Chapters 4, 5 and 6 is obtained by performing a weighting procedure through the queue length, making the users spatially multiplexed and scheduled as

$$\mathbb{K}^*(\mathbf{H}) = \arg \max_{\mathbb{K}} \left(\sum_{k \in \mathbb{K}} Q_k R_k(\mathbf{H}_{\mathbb{K}}, \mathbf{p}(\mathbf{H})) \right) \qquad (7.13)$$

where from previous chapter we know that the optimal power allocation depends on whether we deal with SISO, SIMO or MISO scenarios.

Assume an i.i.d. nature of packet arrivals with an average input rate $\lambda_k = E[a_k(t)]$ and $E[(a_k(t))^2] < \infty$, then combining the queue dynamic in (7.12) with the policy (7.13), the following highest value of the average delay (i.e. the average delay upper bound) is obtained [18, 20]

$$\sum_k \lambda_k \bar{D}_k \leq \frac{TB}{2\varepsilon} \tag{7.14}$$

where ε can be viewed as the minimum distance from λ_k to the boundary of the average spectral efficiency region and

$$B = \max_{\mathbb{R}} \left[\sum_k R_k(\mathbf{H}_{\mathbb{K}}, \mathbf{p}(\mathbf{H}))^2 \right] + \sum_k E\left[\left(\frac{a_k(t)}{T} \right)^2 \right] \tag{7.15}$$

with R reflecting the set of all achievable spectral efficiency values.

7.5 Summary

This chapter reviewed delay performance in wireless systems and more specifically in multi-user wireless systems. Several metrics to account for the delay performance in the system were presented, each standing as an appropriate solution for certain environmental requirements and demands.

The different delay sources in the system were analyzed, where special importance was given to both the access and queueing delay, due to their impact on the final system delay. Related to the access delay, the different multi-user scheduling techniques were presented, each one with a different impact on the access delay metrics.

A number of results were presented for the queueing delay study in wireline and computer networks, but considering the wireless system with its unpredictable and random channel makes the queueing delay study very complicated. The different parameters and delay considerations were discussed within the wireless scenario, to present the state of the art in the study of wireless queueing delay.

References

[1] J. Kim and A. Jamalipour, Traffic management and QoS provisioning in future wireless IP networks, *IEEE Personal Communications*, Vol. 8, Issue No. 10, Oct. 2001.

[2] C. Demichelis and P. Chimento, IP packet delay variation metric for IP performance metrics (IPPM), The Internet Engineering Task Force, IETF-RFC, Nov. 2002.

[3] J.F. Kurose and K.W. Ross, *Computer Networking,* Addison-Wesley Press, 2nd Edition, 2002.

[4] R.A. Berry and R.G. Gallager, Communication over fading channels with delay constraints, *IEEE Trans. Information Theory*, Vol. 48, Issue No. 5, May 2002.

[5] M. Sharif and B. Hassibi, Delay considerations for opportunistic scheduling in broadcast fading channels, *IEEE Trans. Wireless Communications*, Vol. 6, Issue No. 9, Sept. 2007.

[6] H. Fu and D.I. Kim, Analysis of throughput and fairness with downlink scheduling in WCDMA networks, *IEEE Trans. Wireless Communications*, Vol. 5, Issue No. 8, Aug. 2006.

[7] F. Kelly, Charging and rate control for elastic traffic, *European Trans. Telecommunications*, Vol. 8, pp. 33–37, Aug. 1997.

[8] P. Viswanath, D.N. Tse and R. Laroia, Opportunistic beamforming using dumb antennas, *IEEE Trans. Information Theory*, Vol. 48, Issue No. 6, June 2002.

[9] M. Sharif and B. Hassibi, On the capacity of MIMO broadcast channel with partial side information, *IEEE Trans. Information Theory*, Vol. 51, Issue No. 2, Feb. 2005.

[10] M. Kobayashi, G. Caire and D. Gesbert, Transmit diversity versus opportunistic beamforming in data packet mobile downlink transmission, *IEEE Trans. Wireless Communications*, Vol. 55, Issue No. 1, Jan. 2007.

[11] G. Caire and S. Shamai, On the achievable throughput of a multiantenna Gaussian broadcast channel, *IEEE Trans. Information Theory*, Vol. 49, Issue No. 7, July 2003.

[12] M. Sharif and B. Hassibi, A comparison of time-sharing, DPC, and beamforming for MIMO broadcast channels with many users, *IEEE Trans. Communications*, Vol. 55, Issue No. 1, Jan. 2007.

[13] P. Svedman, S.K. Wilson, L.J. Cimini and B. Ottersten, Opportunistic beamforming and scheduling for OFDMA systems, *IEEE Trans. Communications*, Vol. 55, Issue No. 5, May 2007.

[14] M. Kountouris, *Multiuser Multiantenna Systems with Limited Feedback*, Ph.D. dissertation, ENST Paris, France, Dec. 2007.

[15] D. Wu and R. Negi, Utilizing multiuser diversity for efficient support of quality of service over a fading channel, *IEEE Trans. Vehicular Technology*, Vol. 54, Issue No. 5, May 2005.

[16] R. Jain, *The Art of Computer Systems Performance Analysis: Techniques for Experimental Design, Measurement, Simulation, and Modeling*, Wiley-Interscience, New York, April 1991.

[17] L. Kleinrock, *Queueing Systems: Volume 1*, Wiley-Interscience, New York, 1975.

[18] M.J. Neely, E. Modiano and C.E. Rohrs, Dynamic power allocation and routing for time varying wireless networks, *IEEE Journal on Selected Areas in Communications*, Vol. 23, Issue No. 1, Jan. 2005.

[19] N. Zorba and A.I. Pérez-Neira, *CAC for multibeam opportunistic schemes in heterogeneous WiMax systems under QoS constraints*, IEEE Globecom, Washington, USA, Nov. 2007.

[20] H. Boche and M. Wiczanowski, Optimization-theoretic analysis of stability-optimal transmission policy for multiple antenna multiple access channel, *IEEE Trans. Signal Processing*, Vol. 55, Issue No. 6, June 2007.

[21] R. Berry and E. Yeh, Cross-layer wireless resource allocation – fundamental performance limits for wireless fading channels, *IEEE Signal Processing Magazine*, Vol. 21, Issue No. 5, Sept. 2004.

008

8

Orthogonal frequency division multiplexing

010

The focus of this chapter is on OFDM (orthogonal frequency division multiplexing) and its extension to systems with multiple transmit and receive antennas. In order to further exploit the multiuser diversity in the frequency domain OFDMA (orthogonal frequency division multiple access) is also addressed. Next generation broadband wireless standards, e.g. IEEE 802.16e WiMAX and Third Generation Partnership Project – Long Term Evolution (3GPP-LTE), use OFDMA as the preferred physical layer multiple access scheme, especially for downlink, where the increased performance is needed for mobile broadband wireless applications. The problem of allocating time slots, spatial modes, subcarriers, rates and power to the different users in an OFDMA system has been an area of active research in recent years. This chapter summarizes the state of the art in this area and outlines interesting research direction that can be further extended by the reader. Following the same methodology as in previous chapters, the goal is to understand resource allocation in OFDMA and MISO systems, that is, with single antenna terminals.

010

8.1 OFDM and OFDMA

020

Orthogonal frequency division multiplexing (OFDM) is a multicarrier modulation technique that has been chosen as the modulation scheme for several current and next generation broadband communication systems. OFDM is popular especially in broadband wireless communication systems primarily due to its resistance to multipath fading, and its ability to deliver high data rates with reasonable computational complexity. OFDM divides a broadband channel into multiple parallel narrowband subchannels, wherein each subchannel carries a low data rate stream, which sums up to a high data rate transmission.

In an OFDM transceiver, the bits are initially mapped by a bank of quadrature amplitude modulation (QAM) encoders into complex symbols, which are then fed into an inverse fast Fourier transform (IFFT) to ensure the orthogonality of the subchannels. The output is then converted from parallel to serial and modulated onto a carrier to be transmitted over the air through the wireless channel. At the receiver, the reverse operations are performed. In practical wireless channels, channel estimation and equalization is necessary to effectively decode the transmitted information. While the capacity for a single user transmission can be achieved with separate coding for each subcarrier and optimal power waterfilling [1], if multiple users are allowed to transmit simultaneously on the different subcarriers per OFDM symbol (i.e. OFDMA), the problem is not convex anymore. In [2] the authors study the optimal resource allocation and throughput and give a complete characterization of the ergodic capacity region of wideband fading multiple-access channels. In general more than one user will be allocated power on a particular sub-band and, as a result, a multi-user receiver (e.g. using interference cancellation) is required to detect each user's signal because of the non-orthogonal channel access. The problem of allocating time slots, subcarriers, spatial channels, rates and power to the different users in an OFDMA system has therefore been an area of active research. Figure 8.1 shows this idea for an OFDMA downlink system with N carriers and K users that experience different channel gains. This allows the base station, assuming it knows the channel gain information, to allocate resources intelligently in order to maximize some performance metric.

In [3], the authors show that, for the single antenna case, the amount of subcarrier sharing is minimal even in the capacity-achieving case. Thus, assigning only one user to each subcarrier could still achieve transmissions close to capacity, and is essentially the downlink OFDMA transmission scheme. However, near capacity performance can be achieved only when optimal allocation of subcarriers, rates and power is performed. For delay limited capacity, it has been shown that the optimal strategy is successive decoding in each subchannel. However, in systems where successive decoding is required, a user channel estimation error does not only affect the concerned user but has an effect on other users occupying the same band, since this error propagates as users are decoded from the received signal. This motivates the search for orthogonal access strategies. Another advantage of orthogonal signaling is the simplicity of receiver implementation and the reciprocity of the algorithms for both up- and downlinks. It has been shown in [4] that for delay limited capacity, the asymptotically optimal solution (for $K \rightarrow$ inf.) in the case of orthogonal signaling is achieved by letting each user transmit on its best channel and using optimized fractions on each subchannel.

8.1.1 Basic signal model

We consider a single cell OFDMA base station, where we ignore the effect of intercell interference, which we assume to be either absent (sufficient cell separation given the power budget) or simply modeled as additive white Gaussian noise, which increases the noise variance of the signal model. The OFDMA base station has N subcarriers with cyclic prefix, wherein there are K users. The received signal vector for the kth user at the nth OFDM symbol, assuming perfect sample and symbol synchronization, SISO transmission and sufficient cyclic prefix length, is given as

$$\mathbf{y}_k[n] = \mathbf{H}_k[n]\mathbf{P}_k[n]\mathbf{s}_k[n] + \mathbf{w}_k[n] \qquad (8.1)$$

where $\mathbf{y}_k[n]$ and $\mathbf{s}_k[n]$ are the N-length received and transmitted complex-value signal vectors; $\mathbf{P}_k[n] = diag\left\{\sqrt{p_{k,1}[n]}, \ldots, \sqrt{p_{k,N}[n]}\right\}$ is the diagonal gain allocation matrix with $p_{k,m}[n]$ as the power allocated to user k in subcarrier m at time n; $\mathbf{w}_k[n]$ is white Gaussian noise with variance σ_w^2 and zero mean; and

$$\mathbf{H}_k[n] = \text{diag}\{h_{k,1}[n], \ldots, h_{k,K}[n]\} \qquad (8.2)$$

is the diagonal channel response matrix, whose elements are the complex-valued frequency-domain wireless channel fading random processes for the kth user at the nth subcarrier. In the subsequent discussion, we shall drop the index n when the context is clear for notational brevity.

s0030 ## 8.1.2 *Resource allocation*

p0060 The existing approaches that focus on the physical layer transmission optimization for OFDMA can be broadly categorized into two different schemes: (i) minimization of the transmit power subject to minimum quality of service (QoS) parameters for each user, which could be a combination of data rate, bit error rates, delays, etc. [5, 6]; and (ii) maximization of the data rates subject to various QoS and/or resource constraints [7–9]. In any of these categories, the algorithms to find the optimal or near-optimal solution to the problem are too computationally complex for real-time implementation. A popular approach to attain near-optimality is constraint relaxation (see, e.g., [5, 8, 9]). Also, many times the problem is decoupled into two steps [6, 7], wherein the number of subcarriers is initially assigned to each user using a greedy algorithm, followed by the subcarrier assignment step. An interesting approach is the one presented in [10–13], which is based on a Lagrangian relaxation of the power constraints and (possibly) rate constraints, instead of the constraint relaxation proposed previously. This relaxation retains the subcarrier assignment exclusivity constraints, but 'dualizes' the power/rate constraints and incorporates them into the objective function, thereby allowing the dual problem to be solved instead. This dual optimization framework is much less complex, with complexity order $O(NK)$ instead of the typical complexities in the order of $O(NK^2)$, and achieves relative optimality gaps that are less than 10^{-4} (i.e. achieving 99.9999% of the optimal solution). Dual methods work well in OFDMA problems due to the problem structure, i.e. there are typically a lot more subcarriers N than users K. In most OFDMA/multicarrier resource allocation problems, the objective function is separable across the N subcarriers, and the number of constraints is in the order of the number of users K. This makes dual optimization techniques an ideal approach to solving them. Finally, we note that whenever an ergodic rate maximization is carried out, as it has been shown in this book, it results in less complexity per symbol than instantaneous rate maximization, and thus presents an attractive communication performance versus complexity trade-off.

p0070 Concerning the opportunistic schemes, it was shown in [14] that in order to maximize the total capacity, each subcarrier should be allocated to the user with the best gain on it, and the power should be allocated using the water-filling algorithm across the subcarriers. However, no fairness among the users was considered. The same authors extended the problem formulation to consider ergodic rates, which utilizes the temporal dimension when ergodicity of the channel gains is assumed to improve the data rate performance. However, it also suffers from the unfairness problem.

The fairness problem is also addressed in several works, for instance in [7, 15] by ensuring that each user would be able to transmit at a minimum rate, or in [8] by enforcing a notion of max–min fairness, and thus the starvation of some users can be avoided. In [9], prioritization was enforced using a weighted-sum rate maximization. As it has been commented on in previous chapters, by varying the weights for each user's rate, the boundary of the rate region can also be traced out. Several other methods that use various heuristics have also been proposed in order to solve the problem with lower complexity.

Concerning the PHY–MAC cross-layer approach to OFDMA resource allocation, there are fewer works. Instead of using the information-theory rate, they consider a general spectral efficiency cost, which accounts for longer-term throughput and queue state information. In [16], resource allocation that optimizes total packet throughput subject to the user's outage probability constraint was proposed. In [17], throughput maximization coupled with queue load balancing is proposed. More recently, in [18, 19], a cross-layer approach that bridges the gap between the physical (PHY) layer and the media access control (MAC) layer was investigated. It was shown that trade-offs between efficiency and fairness can be realized by maximizing a concave utility function of the user's data rate, instead of maximizing the data rates themselves. We observe that in most of the aforementioned work in physical layer transmit optimization, the formulation and algorithms only consider instantaneous performance metrics. Thus, the temporal dimension is not being exploited when the resource allocation is performed. Although some of the PHY–MAC cross-layer studies consider time-averaged throughput performance, as in [19], their approaches focus more on the effect of the past channel information on fairness, rather than exploiting the temporal variations of the wireless channel directly to improve the overall physical data rate performance, as we have done all along in this book.

p0100   Extending OFDMA to systems with multiple transmit and receive antennas is certainly an interesting problem. Next, MIMO–OFDMA is addressed in order to outline interesting research in this direction.

s0040   ## 8.2 MIMO–OFDMA

p0110   In this case, the following model corresponding to Figure 8.1 has to be considered

$$\mathbf{y}^m[n] = \mathbf{H}^m[n]\mathbf{P}^m[n]\mathbf{s}^m[n] + \mathbf{w}^m[n] \tag{8.3}$$

where, differently to model (8.1), $\mathbf{H}^m[n]$ is the $N \times Q$ complex flat-fading channel matrix at the mth subcarrier, the kth row of which, $\mathbf{H}_k(m,n)$ contains the $1 \times Q$ vector of the channel gains for the kth user at the mth subcarrier. Also, in case of transmit beamforming, $\mathbf{s}^m[n] = \mathbf{T}^m[n]\mathbf{x}^m[n]$, where $\mathbf{T}^m[n]$ is a matrix of dimension $Q \times K$ that gathers the transmit beamvectors and $\mathbf{x}^m(n)$ contains the information symbols. The mth subchannel signal SINR of the kth receiver can be denoted as

$$SINR_k(m, n) = \frac{p_k(m, n)\left|\mathbf{t}_k^H(m, n).\mathbf{h}_k(m, n)\right|^2}{\displaystyle\sum_{\substack{n=1 \\ n \neq k}}^{K} p_n(m, n)\left|\mathbf{t}_n^H(m, n).\mathbf{h}_k(m, n)\right|^2 + \sigma^2} \tag{8.4}$$

The sum rate that can be achieved by the kth user is

$$C_k = \sum_{m=1}^{N} \log_2(1 + SINR_k(m, n)) \tag{8.5}$$

and the system throughput of all users is

$$C = \sum_{m=1}^{M} \sum_{k=1}^{K} \log_2(1 + SINR_k(m, n)) \tag{8.6}$$

The algorithms proposed in this book can be extended to transmit/ receive beamforming [20] or spatial multiplexing as long as the objective function is separable across the subcarriers; one example

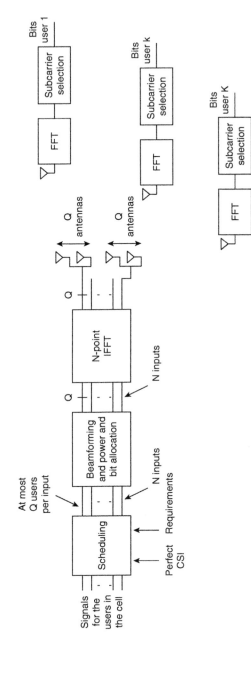

Figure 8.1 *Block diagram of OFDMA downlink system. With perfect channel state information (CSI), the access point (AP) clusters the users into groups, i.e. the scheduling task. For each group at each of the N subcarriers, the AP performs a specific spatial precoder, and the corresponding power and bit allocation. The N inputs of most Q users feed the IFFT block of the OFDM system. At the receivers, terminals shall demodulate only the signals at their subcarriers. It is assumed that this information is sent by the AP through a broadcast channel.*

is zero forcing (ZF). Therefore, a ZF transmit/receive beamforming, the signal model in the last chapter applies also to OFDMA if an additional index to indicate the subcarrier number. After using the ZF beamforming, the channel of the kth user is transformed into parallel channels and $SNR_k(m,n)$ is the one present in equations (8.5) and (8.6) instead of $SNIR_k$. In [21] several scheduling strategies for a ZF beamformer are studied within the context of OFDMA, ranging from the optimal one with a highest complexity towards simple suboptimal solutions. On top of the transmit beamforming, the access point performs the corresponding spatial power and bit allocation. Bit allocation is derived in a power minimization (or bit rate maximization) under QoS constraints. See Chapter 5 of [21] for a review of different bit allocation mechanisms in OFDMA. However, the use of a ZF transmit beamforming forces the modification of the traditional single antenna bit loading schemes, basically because the channels from the users change depending on those scheduled. This topic is addressed in [21] using a modified capacity formula that takes into account the gap due to the bit error rate.

p0120 Different from the ZF beamforming, other low complextity MISO–OFDMA strategies have been studied in the literature that are based on opportunistic beamforming design [22]. In [23], instead of directly extending the single carrier design to OFDM, since such a system consists of parallel flat fading channels, the authors propose a clustered beamforming design (i.e. following the same line of thought as in [21]). By properly grouping subcarriers not only complexity is reduced but frequency variability is induced in a favorable way. The results are given for the 3GPP long-term evolution (LTE) of UMTS, which aims to offer 100 Mbit/s in the downlink by using OFDMA combined with MIMO technologies in an up to 20 MHz bandwidth. As a first step towards a cross-layer approach for the combination of beamforming and scheduling in MIMO–OFDMA systems, in [24], the authors present a combination of a self-organized beamforming approach with opportunistic scheduling algorithms for OFDMA. In the considered system, different beams are formed and grouped together. The space–time frequency resources are then allocated to mobile terminals by intelligent scheduling methods such that mobile terminals are assigned resources when a high instantaneous SINR is observed. The work also considers an MAC frame structure to take into account the enhanced MIMO–OFDM scheduling. This scheduling is designed according to the achievable goodput and fairness (e.g. goodput maximization scheduling, proportional fairness scheduling). In contrast to [21, 23], this work is done under a fully cross-layer approach; however, the performance of the system proposal is only quantitatively assessed in terms of system throughput leaving space for a more analytical study.

In a more general case of using 'spatial waterfilling' in MIMO–OFDMA, theoretically, we could perform a singular value decomposition (SVD) per subcarrier; we would then have min(Nr, Q) 'spatial gains' per subcarrier. Thus, instead of the N SNR values in the SISO–OFDMA case, we have N min(Nr, Q) SNR values, and we could reuse the existing methods developed for the SISO–OFDMA, since we could similarly use 'multi-level waterfilling' power allocation across all spatial gains of all subcarriers. However, this is a theoretical approach as it lacks the multiple access component because it only deals with a point to point MIMO transmission, which does not take into account that the receivers are not collocated in the same terminal. Note also that when looking into not only the physical layer, but also the link layer, feedback load is especially critical in multi-user MIMO systems because of its higher dimensionality. In [25] this special issue focuses on such transmission systems with low amounts of feedback and provides an overview of the state of the art.

8.3 Summary

Developing efficient algorithms for MIMO–OFDMA is an interesting avenue for further research. In order to design the algorithms, this chapter has reinforced the main ideas that have been exposed in this book: (i) cross-layer spectral efficiency offers more realistic results than information-theoretic capacity; (ii) ergodic instead of instantaneous optimization makes the problem less complex and also allows time allocation; and (iii) dual methods work well in OFDMA problems due to the problem structure, i.e. there are typically a lot more subcarriers N than users K. In addition, recently [26] has provided a new interpretation of the water-filling operator, for the general MIMO multi-user case, as a matrix projection, and possible future work is to extend it to OFDMA.

References

[1] D. Tse and P. Viswanath, *Fundamentals of Wireless Communication*, Cambridge University Press, May 2005.

[2] D.N.C. Tse and S.D. Hanly, Multiaccess fading channels. I: Polymatroid structure, optimal resource allocation and throughput capacities, *IEEE Trans. Information Theory*, Vol. 44, pp. 2796–2815, Nov. 1998.

[3] A. Goldsmith and M. Effros, The capacity region of broadcast channels with intersymbol interference and colored Gaussian noise, *IEEE Trans. Information Theory*, Vol. 47, Issue No. 1, pp. 219–240, Jan. 2001.

[4] G. Caire, R. Müller and R. Knopp, Multiuser diversity in delay limited cellular wideband systems, *IEEE Trans. Information Theory*, Vol. 47, Issue No. 3, March 2001.

[5] C.Y. Wong, R. Cheng, K. Lataief and R. Murch, Multiuser OFDM with adaptive subcarrier, bit, and power allocation, *IEEE J. Select. Areas Comm.*, Vol. 17, Issue No. 10, pp. 1747–1758, Oct. 1999.

[6] D. Kivanc, G. Li and H. Liu, Computationally efficient bandwidth allocation and power control for OFDMA, *IEEE Trans. Wireless Communication*, Vol. 2, Issue No. 6, pp. 1150–1158, Nov. 2003.

[7] H. Yin and H. Liu, An efficient multiuser loading algorithm for OFDM-based broadband wireless systems, *Proc. IEEE Global Telecommunications Conference*, Vol. 1, pp. 103–107, Dec. 2000.

[8] W. Rhee and J.M. Cioffi, Increase in capacity of multiuser OFDM system using dynamic subchannel allocation, *Proc. IEEE Vehicular Technology Conference*, pp. 1085–1089, Tokyo, Japan, May 2000.

[9] L. Hoo, B. Halder, J. Tellado and J. Cioffi, Multiuser transmit optimization for multicarrier broadcast channels: asymptotic FDMA capacity region and algorithms, *IEEE Trans. Wireless Communication*, Vol. 52, Issue No. 6, pp. 922–930, June 2004.

[10] K. Seong, M. Mohseni and J. Cioffi, Optimal resource allocation for OFDMA downlink systems, *Proc. IEEE International Symposium on Information Theory*, pp. 1394–1398, Seattle, WA, July 2006.

[11] Y. Yu, X. Wang and G.B. Giannakis, Channel-adaptive congestion control and OFDMA scheduling for hybrid wireline-wireless networks, *IEEE Trans. Wireless Communication*.

[12] W. Yu and R. Lui, Dual methods for nonconvex spectrum optimization of multicarrier systems, *IEEE Trans. Wireless Communication*, Vol. 54, Issue No. 7, pp. 1310–1322, July 2006.

[13] G. Wunder and T. Miche, Multiuser OFDMA optimization: algorithms and duality gap analysis, Workshop on Smart Antenna, Darmstadt, Feb. 2008.

[14] J. Jang and K.B. Lee, Transmit power adaptation for multiuser OFDM systems, *IEEE J. Select. Areas Comm.*, Vol. 21, pp. 171–178, Feb. 2003.

[15] Y. Zhang and K. Letaief, Multiuser subcarrier and bit allocation along with adaptive cell selection for OFDM transmission, *Proc. IEEE International Conference on Communications*, Vol. 2, pp. 861–865, April 2002.

[16] G. Li and H. Liu, Dynamic resource allocation with finite buffer constraint in broadband OFDMA networks, *Proc. IEEE Wireless Communications and Networking Conference*, Vol. 2, pp. 1037–1042, March 2003.

[17] S. Kittipiyakul and T. Javidi, Subcarrier allocation in OFDMA systems: beyond water-filling, *Proc. IEEE Asilomar Conference on Signals, Systems and Computers*, Vol. 1, pp. 334–338, Nov. 2004.

[18] G. Song and Y. Li, Cross-layer optimization for OFDM wireless networks. Part I: Theoretical framework, *IEEE Trans. Wireless Communication*, Vol. 4, Issue No. 2, pp. 614–624, March 2005.

[19] G. Song and Y. Li, Cross-layer optimization for OFDM wireless networks. Part II: Algorithm development, *IEEE Trans. Wireless Communication*, Vol. 4, Issue No. 2, pp. 625–634, March 2005.

[20] M. Olfat, K.J.R. Liu and F. Rashid-Farrokhi, Low complexity adaptive beamforming and power allocation for OFDM

over wireless networks, *Proc. IEEE Int. Conf. Commun.'99*, Vol. 1, pp. 523–527.

[21] D. Bartolome, *Fairness analysis of wireless beamforming schedulers,* Ph.D. dissertation, UPC Barcelona, Spain, Nov. 2004.

[22] N. Zorba, Multibeam opportunistic downlink beamforming in wireless communication systems, Ph.D. dissertation, UPC Barcelona, Spain, Dec. 2007.

[23] P. Svedman, *Cross-layer aspects in OFDMA systems*, Ph.D. dissertation, KTH-Stockholm, Sweden 2007.

[24] H. Rohling, R. Grünheid and A. Tassoudji, Beamforming and scheduling in a cellular OFDM system, Workshop on Smart Antenna, Darmstadt, Feb. 2008.

[25] MIMO transmission with limited feedback, *EURASIP Journal on Advances in Signal Processing*, Dec. 2007, ISSN 0165-1684, Ed. by M. Rupp, A. Pérez-Neira, C. Mecklenbauker and D. Gesbert.

[26] G. Scutari, D.P. Palomar, and S. Barbarossa, Competitive design of multiuser MIMO interference systems based on game theory: a unified approach, *ICASS, 2008*, Las Vegas, USA.

Index

Printed and bound by CPI Group (UK) Ltd, Croydon, CR0 4YY

03/10/2024

01040418-0020